GCSE

Mathematics

Practice Exam Papers
Answer Book

Higher Tier

Exam Set MHP43

These practice papers won't make you better at maths

... but they will show you what you **can** do, and what you **can't** do.

These are GCSE papers, just like you'll get in your exams — so they'll tell you what you need to **work at** if you want to do **better** on the day.

Do an exam, **mark it** and look at what you **got wrong**.
That's the stuff you need to learn.

Go away, **learn** those tricky bits, then **do the same exam again**. If you're **still** getting questions wrong, you'll have to do even **more practice** and **keep testing** yourself until you keep getting **all** the questions right.

It doesn't sound like a lot of **fun**, but it **really will help**.

The three big ways to improve your score

1) **Answer all these exams**
   These practice papers contain all the types of question that have come up year after year in GCSE exams. If you can do all these, you should be able to do all the questions in your exams.

2) **Keep practising the things you get wrong**
   The whole point of a practice exam is to find out what you don't know*. So every time you get a question wrong, revise that subject then have another crack at it.

   *Use the mark scheme in this booklet to help you see where you dropped your marks.

3) **Don't throw away easy marks**
   Always answer the question the way it's asked — if it asks for units, use the right ones. Always double-check your answer and don't make silly mistakes — obvious really.

Remember: the fewer marks you lose, the more marks you get.

Working out your Grade

- Do a complete exam (paper 1 and paper 2).

- Use the answers and mark scheme in this booklet to mark each exam paper. The marks are all out of 100, so they're already percentages.

- Find your average percentage for the whole exam (both papers).

- Look it up in this table to see what grade you got.

Average %	84 +	69 – 83	46 – 68	32 – 45	20 – 31	under 20
Grade	A*	A	B	C	D	E

Stick your marks in here so you can see how you're doing

		Paper 1 %	Paper 2 %	Average %	Grade
SET 1	First go				
	Second go				
	Third go				
SET 2	First go				
	Second go				
	Third go				
SET 3	First go				
	Second go				
	Third go				

Important!

Any grade you get on one of these practice papers is **no guarantee** of getting that in the real exam — **but** it's a pretty good guide.

Published by CGP.

Contributors:
Cath Brown, Katherine Craig, Rosie Hanson,
David Hickinson, Rosie McCurrie, Rosemary Rogers,
Jane Towle, Julie Wakeling, Sarah Williams.

Many thanks to Katie Braid and Vicky Daniel
for the proofreading.

Groovy website: www.cgpbooks.co.uk
Jolly bits of clipart from CorelDRAW®
Printed by Elanders Ltd, Newcastle upon Tyne.

Psst… photocopying these practice papers isn't allowed, even
if you've got a CLA licence. Luckily, it's dead cheap, easy and
quick to order more copies from CGP — just call us
on 0870 750 1242. Phew!

These answers and mark schemes will show you exactly how to do each question and where you get the marks. If you've got to the correct answer by a different method, you can still award yourself full marks, as long as your answer and working are absolutely clear...

<u>Set 1 Paper 1 — Non-calculator</u>

1 Round off the numbers $\dfrac{30\,000 \times 0.1}{3 + 7} = \dfrac{3000}{10} = \mathbf{300}$

[3 marks available — 1 mark for rounding the numbers in the question (available for rounding two or more of them correctly); 1 mark for one correct calculation; 1 mark for the correct answer.]

Pick sensible numbers when ROUNDING OFF so you can do the sums.

2 a) $2xt = 4a - 4$, $t = \dfrac{4a - 4}{2x}$, $t = \dfrac{2a - 2}{x}$ OR $t = \dfrac{2a}{x} - \dfrac{2}{x}$

[2 marks available — 1 mark for showing '2xt = 4a – 4'; 1 mark for the correct final answer.]

b) $t = \dfrac{2a - 2}{x}$, $t = \dfrac{(2 \times 10) - 2}{2}$, $t = \dfrac{18}{2}$, $t = \mathbf{9}$

[2 marks available — 1 method mark for substituting values for a and x into answer from part a); 1 mark for the correct final answer.]

If you're not sure whether your answer to part a) is right, put the values into the original equation and see if you get the same answer. So, 2(2)t + 4 = 4(10), 4t + 4 = 40, 4t = 36, t = 9.

3 a) 2 cm = 1 m.
5.5 m = 5.5 × 2 = **11 cm**.
4.2 m = 4.2 × 2 = **8.4 cm**.
[2 marks available — 1 mark for each correct answer.]

b)
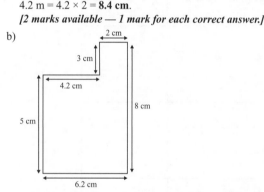

[3 marks available — 1 mark for every two lengths drawn and labelled correctly.]

4 a)
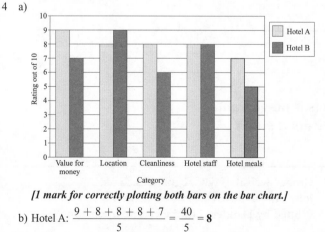

[1 mark for correctly plotting both bars on the bar chart.]

b) Hotel A: $\dfrac{9 + 8 + 8 + 8 + 7}{5} = \dfrac{40}{5} = \mathbf{8}$

Hotel B: $\dfrac{7 + 9 + 6 + 8 + 5}{5} = \dfrac{35}{5} = \mathbf{7}$

Jinny and Tim should stay at Hotel **A** because it scores a higher mean rating.
[3 marks available — 2 marks for correctly calculating the mean for both hotels; 1 mark for choosing Hotel A.]

5 a) 360/36 = 10° per pupil. Angle for grade D or below measures 100°. So 100/10 = **10 pupils** got grade D or below.
[2 marks available — 1 mark for showing 10° on the pie chart is equivalent to one pupil; 1 mark for the correct final answer.]

b) Donya is incorrect. E.g. you don't know how many people are in Class 10Y. Although the angle in the pie chart is bigger for 10H, that only tells you about the fraction/proportion/percentage of each class getting the grade. If there are fewer people in 10H than in 10Y, then although it's a bigger fraction it could mean fewer people got a grade A.
[4 marks available — 1 mark for stating that Donya is incorrect; 1 mark for explaining that the pie chart shows the fraction/ proportion/percentage of people getting the grade, not the number; 1 mark for explaining that if class 10H is smaller, this could mean Donya is wrong; 1 mark for use of good English, with no spelling mistakes or poor grammar.]

Stop dreaming of steaks and kidneys, and get practising those PIE CHARTS.

6 a) 1. The question is asked in a way that is expecting a particular answer / it is a leading question — it would be better to say something like: "What do you think of school meals?".
2. Only allowing a Yes/No answer may not collect accurate information, e.g. some children might want to say "sometimes" or say which meals they liked and which they didn't.
[2 communication marks available — 1 mark for criticising the wording of the questionnaire; 1 mark for criticising the limited options for answers. A suitable explanation must be given to get the mark in each case.]

b) E.g. it's a good method because the children have just eaten in the canteen, so they won't forget about what they think of the food. It's a bad method because the children who don't eat in the canteen won't be asked as they won't be leaving the canteen. Having to give the questionnaires to the Headteacher could make some students biased, as they won't want to give their true opinion.
[2 communication marks available — 1 mark each for any of the above 3 points or an equivalent answer.]

7 a) $3 \times 10^{200} \times 2 \times 10^{100} = 6 \times 10^{(200 + 100)} = \mathbf{6 \times 10^{300}}$
[2 marks available — 1 mark for '6 × 10^{some power}'; 1 mark for '10^{300}']

b) $3 \times 10^5 + 2 \times 10^4 = 30 \times 10^4 + 2 \times 10^4 = 32 \times 10^4 = \mathbf{3.2 \times 10^5}$
[2 marks available — 1 mark for '3.2 × 10^{some power}' or '32 × 10^4'; 1 mark for '10^5'.]

c) $(3 \times 10^{20}) \div (2 \times 10^{30}) = 3/2 \times 10^{(20 - 30)} = \mathbf{1.5 \times 10^{-10}}$
[2 marks available — 1 mark for '1.5 × 10^{some power}'; 1 mark for '10^{-10}'.]

8 a) Let x = P(yellow sweet)
So $1/7 + x + 2x = 1$
$3x = 1 - 1/7$
$3x = 6/7$
$x = 6/7 \div 3 = \mathbf{2/7}$

[3 marks available — 1 mark for '1/7 + x + 2x = 1' or equivalent formula; 1 mark for solving formula for x (or equivalent); 1 mark for the correct final answer.]

b) There must be 7 sweets in the bag. You can only have whole numbers of sweets, so 1/7th of the total number of sweets must be a whole number. 7 is the only number under 10 that this is true for.
[2 marks available — 1 mark for correct reason; 1 communication mark for a clear explanation.]

9 Angle EDB = 20°, as alternate angles are equal.
Angle BED = 20°, as BDE is isosceles so 2 angles are equal.
So angle ABE = **20°**, as alternate angles are equal.
Or, angle EBD = 140°, as angles in a triangle add up to 180°.
So angle ABE = **20°**, as angles on a straight line add up to 180°.
[3 marks available — 1 mark for angle EDB; 1 mark for angle BED; 1 mark for the correct final answer.]

"It looks the same size as the other angle" won't get you the marks here — you need to know all the ANGLE RULES.

10 a)

x	0	1	2	3	4	5
$y = \dfrac{60}{x+1}$	60	30	20	15	12	10

[2 marks available — lose 1 mark per mistake.]

b)

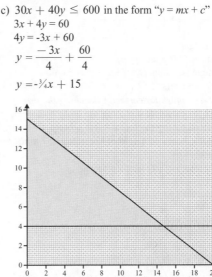

[2 marks available — 1 mark for all points plotted correctly (errors may be carried forward from part a); 1 mark for joining the points with a smooth curve.]

c) i) $y = 10.9$ (accept values between 10.5 and 11.0).

[1 mark available; also accept a correct value read off an incorrect graph resulting from part a) or b).]

ii) $x = 1.4$ (accept values between 1.3 and 1.5).

[1 mark available; also accept a correct value read off an incorrect graph resulting from part a) or b).]

Your values are more likely to be correct if you draw your curve nice and smooth.

11 a) i) $44 \div 220 \times 100\% = \textbf{20\%}$
[1 mark]

ii) 20% of £310 = £62
£310 − £62 = **£248**
[2 marks available — 1 mark for calculating 20% of £310; 1 mark for an answer of £248. (Follow-through marks available if part a) i) is wrong.)]

b) £130 is 20%, which is 1/5 of the total amount he was taxed on.
£130 × 5 = £650
£650 + £420 = **£1070**.
[2 marks available — 1 mark for finding his taxable income '£650'; 1 mark for the correct final answer. (Follow-through marks available if part (a) i) is wrong.)]

12   P(Head) = 1/2; P(Tail) = 1/2
P(3 Heads) = 1/2 × 1/2 × 1/2 = 1/8
P(3 Tails) = 1/2 × 1/2 × 1/2 = 1/8
P(3 Heads or 3 Tails) = 1/8 + 1/8 (OR rule) = 1/4
Kate has 1/4 chance of winning.
Darren has 1 − 1/4 = 3/4 chance of winning,
so **Darren** has the best chance of winning.
[5 marks available — 1 method mark for a suitable way to calculate P(3 Heads or 3 Tails), e.g. by using multiplication rule, listing outcomes or using a tree diagram; 1 mark for getting either P(3 Heads) or P(3 Tails) = 1/8; 1 mark for P(Kate wins) = 1/4; 1 mark for P(Darren wins) = 3/4; 1 communication mark for concluding Darren has a better chance of winning with correct reasoning.]

Another PROBABILITY question, what are the chances of that?

13 a) Perimeter of lawn = 2 × radius + arc length.
2 × radius = 2 × 3 m = 6 m.

Arc length = $\dfrac{angle\ of\ sector}{360} \times \pi d = \dfrac{150}{360} \times 6\pi$

$= \dfrac{15}{36} \times 6\pi = \dfrac{5}{12} \times 6\pi = \dfrac{30\pi}{12}$, so perimeter = $\dfrac{\textbf{5}\pi}{\textbf{2}} + \textbf{6 m}$.

[3 marks available — 1 method mark for attempting to find the arc length by considering the fraction of the circle's circumference; 1 mark for the correct arc length; 1 mark for the correct final answer.]

b) Area of sector = $\dfrac{angle\ of\ sector}{360} \times \pi r^2$

$= \dfrac{150}{360} \times \pi(3^2) = \dfrac{15}{36} \times 9\pi = \dfrac{5}{12} \times 9\pi = \dfrac{45\pi}{12} = \dfrac{15\pi}{4}$

So cost = $\dfrac{15\pi}{4} \times 28 = \dfrac{15}{4} \times \dfrac{22}{7} \times \dfrac{28}{1} = 15 \times 22 = \textbf{£330}$

[3 marks available — 1 method mark for attempting to find the area of the sector by considering the sector as a fraction of a circle; 1 mark for finding the correct area of sector; 1 mark for the correct final answer.]

Tricky calculations here, so take your time and check your answers.

14 a) The total cost of the cola = $30x$
The total cost of the lemonade = $40y$
So altogether Danny pays $30x + 40y$
He has only £6 to spend, which is 600p, so that means his total spend on cola and lemonade can't be more than 600.
So $\mathbf{30x + 40y \leq 600}$.
[1 mark for the correct answer.]

b) Danny must buy at least 4 cans of lemonade, so $\mathbf{y \geq 4}$.
[1 mark.]

c) $30x + 40y \leq 600$ in the form "$y = mx + c$"
$3x + 4y = 60$
$4y = -3x + 60$
$y = \dfrac{-3x}{4} + \dfrac{60}{4}$
$y = -\tfrac{3}{4}x + 15$

[5 marks available — 1 mark for drawing a line at y = 4; 1 mark for rearranging '30x + 40y ≤ 600' into the form 'y = mx + c'; 1 mark for drawing the line 30x + 40y = 600 correctly; 1 mark for drawing both lines with solid lines; 1 mark for shading the correct region.]

d) $30x + 40y \leq 600$, $x + y = 18$ and $y \geq 4$.
So the possibilities that fall in the shaded region on the graph are:
12 cans of cola and 6 cans of lemonade;
OR 13 cans of cola and 5 cans of lemonade;
OR 14 cans of cola and 4 cans of lemonade.
[1 mark available for any one of the above answers.]

15 a) Exterior angle = 180 − 150 = 30°
Exterior angle = 360 ÷ no. of sides
So no. of sides = 360 ÷ 30 = **12 sides**.
[3 marks available — 1 mark for identifying the exterior angle; 1 mark for attempt to use '360/number of sides'; 1 mark for the correct final answer.]

b) As the interior angle is not a factor of 360°, you cannot make up a complete 360° using just these polygons — there will always be gaps between the shapes.
[2 marks available — 1 mark for the idea that you need the interior angle to be a factor of 360°; 1 mark for the idea that using this polygon will leave gaps between the shapes.]

16 a) $2^3 \times (2^x)^4 = 2^3 \times 2^{4x} = 2^{3+4x}$.
[2 marks available — 1 mark for '2^{4x}'; 1 mark for the correct final answer.]

You can think of $(2^x)^4$ as $2^x \times 2^x \times 2^x \times 2^x$ and the add the powers... but it's much easier to LEARN those power rules.

b) $2^{5+3x} = 2^2$
$5 + 3x = 2$, so $x = $ **-1**.

[2 marks available — 1 mark for '$5 + 3x = 2$'; 1 mark for the correct final answer.]

17 a) The gradient of line given is -2.
So the gradient of perpendicular line is $-1/-2 = ½$.
y intercept = 8
So the equation of the perpendicular line is $y = ½ x + 8$.
[2 marks available — 1 mark for 'gradient = ½'; 1 mark for the correct final answer.]

b)

Area = ½ × base × height.
Using the sketch (turned 90° clockwise),
the base of the triangle is 5 units and the height is 2 units.
So area = ½ × 5 × 2 = **5 units**.
[5 marks available — 1 mark for correctly sketching the line $y = ½ x + 8$; 1 mark for correctly sketching the line $y = -2x + 3$; 1 mark for correctly sketching the line $x = 0$ to join up the triangle; 1 mark for finding the base and height of the triangle; 1 mark for the correct final answer.]

You should really learn how to find the area of a triangle — there might be a cheeky question about triangles in the exam.

18

a) Any changes outside the f(x) brackets affect the y values.
So for f(x) + 1, all the y values increase by 1 (see graph).
[2 marks available — 1 mark for a clear attempt at moving the graph upwards whilst keeping its shape the same; 1 mark for a correct, accurate line.]

b) For $y = 1.5$f(x), all the y values are multiplied by 1.5.
[2 marks available — 1 mark for a clear attempt to stretch the graph in the y-direction; 1 mark for a correct, accurate line.]

19 $\sqrt{32} = \sqrt{16} \times \sqrt{2} = 4\sqrt{2}$
$\sqrt{8} = \sqrt{4} \times \sqrt{2} = 2\sqrt{2}$
$\sqrt{27} = \sqrt{9} \times \sqrt{3} = 3\sqrt{3}$
So $(\sqrt{32} + \sqrt{3})(\sqrt{8} - \sqrt{27}) = (4\sqrt{2} + \sqrt{3})(2\sqrt{2} - 3\sqrt{3})$
Multiply out: $8\sqrt{2}\sqrt{2} - 12\sqrt{2}\sqrt{3} + 2\sqrt{2}\sqrt{3} - 3\sqrt{3}\sqrt{3}$

Combine square roots: $8(2) - 12\sqrt{6} + 2\sqrt{6} - 3(3)$
$= 16 - 10\sqrt{6} - 9 = \mathbf{7 - 10\sqrt{6}}$
[5 marks available — 1 method mark for attempting to simplify surds (this mark may also be obtained by attempting to simplify an answer after multiplying out); 1 mark for all simplifications correct; 1 method mark for multiplying out the brackets; 1 mark for showing knowledge that $\sqrt{x} \times \sqrt{x} = x$; 1 mark for the correct final answer.]

When combining square roots, remember that $\sqrt{2}\sqrt{2} = 2$ (easy).

20 Represent an even number algebraically as $2n$, where n is a whole number (since an even number is 2 × something).
So an odd number can be represented as $2n + 1$ (or $2n - 1$).
Squaring an odd number = $(2n + 1)^2 = (2n + 1)(2n + 1)$
$= 4n^2 + 4n + 1$.
Subtracting 1 gives $4n^2 + 4n$.
This can be written as $4(n^2 + n)$.
Since n is a whole number, $n^2 + n$ are whole numbers too.
So the result is 4 × a whole number, so it is divisible by 4.
[5 marks available — 1 mark for representing an odd number as '$2n + 1$' (or use '$2n - 1$' throughout); 1 mark for squaring correctly; 1 mark for '$4(n^2 + n)$'; 1 communication mark for explaining how this shows the result is divisible by 4; 1 communication mark for a clear, ordered answer.]

Don't be put off by being asked to use ALGEBRA — it just means you shove a letter in there to represent what you don't know.

Set 1 Paper 2 — Calculator

1 Contract A: Total cost is £20 for the contract, plus the cost of the extra calls. He will have 100 minutes of extra calls, which will cost $100 \times 10p = £10$. So the total cost is £30.
Contract B: Total cost is £25 for the contract, plus the cost of the extra texts. He will have 100 extra texts which will cost $100 \times 10p = £10$. So the total cost is £35.
Therefore he should choose **Contract A**.
[3 marks available — 1 mark for finding the total cost for Contract A; 1 mark for finding the total cost for Contract B; 1 mark for choosing Contract A.]

2 Volume of draft excluder = $\pi r^2 \times l$
$= \pi \times (0.16/2)^2 \times 1.2$
$= 0.024127431...$ m^3

Filling = 0.03 m^3
So she should buy **1 ft^3** bag.
[3 marks available — 1 method mark for correctly substituting values into the volume formula; 1 mark for converting the measurement into m; 1 mark for the correct final answer.]

3 a) Jenny's estimate is $220 \div 365 = 0.603 = 60.3\%$,
so Jenny's estimate is higher than 41% from Danzeela's data.
[2 marks available — 1 method mark for attempting to work out '$220 \div 365$'; 1 mark for a correct answer and a comparison involving writing both estimates as decimals OR both as percentages OR both as fractions.]

b) Danzeela's estimate is more reliable. This is because her data comes from a larger sample size. / Jenny's estimate may be less accurate because last year may have been unusual so contain anomalous data. / Jenny's data is only for where she lives and not the whole of the UK, so Danzeela's estimate is more reliable.
[3 marks available — 1 mark for identifying that Danzeela's data is more reliable; 1 mark for a clear explanation similar to one of the above explanations; 1 communication mark for use of specialist vocabulary, e.g. 'sample size' or 'anomalous'.]

All this STATISTICS stuff really isn't too bad.

4 a) $(a-3)(a+4) = a^2 + 4a - 3a - 12$
$= \mathbf{a^2 + a - 12}$
[2 marks available — 2 marks for getting the correct final answer;
1 mark for getting at least 2 correct terms from 'a² + 4a – 3a – 12'.]

Don't forget — you get four terms with double brackets...

b) $3xy - 9x^2y = \mathbf{3xy\,(1 - 3x)}$
[2 marks available — 1 mark for '3xy'; 1 mark for '(1 – 3x)'.
Give 1 mark for an incomplete factorisation, such as
'3x(y – 3xy)', '3y(x – 3x²)' or 'xy(3 – 9x)'.]

c) $x^2 + 6x + 8$
$= \mathbf{(x+4)(x+2)}$
[2 marks available — 1 mark for '(x + 4)'; 1 mark for '(x + 2)'.]

Check you get x² + 6x + 8 when you expand your answer.
If you wrote (x + 8)(x + 1) then you'll get the x² + 8 bit,
but you'll also get 8x + x = 9x... wrong... try again...

5 a) The sequence goes up by 3 each time.
20 and **23**
[1 mark]

b) $3n$ because the sequence goes up in 3s.
$+2$ because $3n$ on its own gives 3, 6, 9, 12... so we have to add 2
to get the sequence given in the question.
So n^{th} term $= \mathbf{3n + 2}$.
[2 marks available — 1 mark for '3n'; 1 mark for '+2'.]

c) $3n + 2 > 100$
$3n > 98$
$n > 98/3 = 32.666...$
So, the first whole number term with a value larger than 100 is the
33rd term (or 101).
[3 marks available — 1 method mark for writing '3n + 2 >100' or
'3n + 2 = 101' or for using a logical method, such as finding how
many more times 3 has to be added on;
1 mark for 'n > 98/3'; 1 mark for the correct final answer.]

6 a)

3	5			
4	0	0	5	
5	0			
6	0	0	0	5
7	0			
8	0	5	5	
9	0	0	0	
10	0			
11				
12	0	0	5	

Key 3 | 5 means 35.

[3 marks available — 1 mark for all entries correct;
1 mark for all entries in the correct ascending order;
1 mark for a correct key.]

b) Year 7 median = halfway between 70 and 80 = 75 minutes.
Year 7 median plus 30 min revision = 75 + 30 = **105 minutes**.
[2 marks available — 1 mark for finding the correct Year 7 median;
1 mark for correctly working out the new median.]

7 Total price for website A: (1.2 × £80) + £6 = £96 + £6 = £102
(or: 20/100 × £80 = £16, then £80 + £16 + £6 = £102)

Total price for website B: £100

Total price for website C: 0.8 × £120 = £96
(or: 20/100 × £120 = £24, then £120 – £24 = £96)

So he should buy it from **website C**.
[3 marks available — 1 mark for getting the correct answer for
website A; 1 mark for getting the correct answer for website C;
1 mark for the correct final answer.]

8 a) $4q + 9 = 1$
$4q = -8$
$q = -2$
[2 marks available — 1 mark for showing '4q = -8';
1 mark for the correct final answer.]

b) $y^3 - 18 = 36 - y^3$
$y^3 = 54 - y^3$
$2y^3 = 54$
$y^3 = 27$
$\mathbf{y = 3}$
[3 marks available — 1 mark for showing '2y³ = 54'; 1 mark
for showing 'y³ = 27'; 1 mark for the correct final answer.]
SOLVING EQUATIONS — oh, life couldn't be more fun...

9 a) Total probability = 1.
10% is 0.1 as a decimal, and ¼ is 0.25.
So P(Shannon visits her friend) = 1 – 0.1 – 0.25 = **0.65**
[2 marks available — 1 mark for '1 – 0.1 – 0.25';
1 mark for the correct answer expressed as a decimal.]

b) AND rule, so multiply: 0.1 × 0.1 = **0.01** (also accept **1/100**)
[2 marks available — 1 mark for spotting the AND rule
and attempting to multiply; 1 mark for correct final answer.]

c) She went to the cinema ¼ of the time so 15 weeks is ¼ of the number
of weeks. So total number of weeks = 15 × 4 = **60 weeks**.
[1 mark available for correct answer only.]

10 a) The graph shows a positive correlation between the distance she jogs
and her weight loss.
[2 marks available — 1 mark for 'positive';
1 communication mark for using the word 'correlation'.]

b)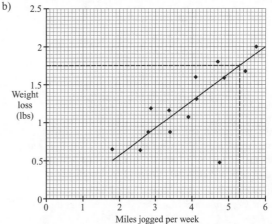

[1 mark for a line of best fit similar to the one shown above.]

c) From the graph, 5.3 miles.
[1 mark for correctly reading a value off the graph
(accept answers in the range 5-6 miles).]

Starting to lose the plot? Have another look at SCATTER GRAPHS.

11 Split the floor up into a rectangle and a semicircle.
Area of rectangle = 3.5 × 4.5 = 15.75 m²
Area of semicircle = ½ π r² = ½ π 1² = 1.57 m² (to 2 d.p.)
Total area = 17.32...m²
Total volume of concrete = 17.32... × 0.1 = 1.732...m³
Total cost = 1.732... × £80 = **£139**
[4 marks available — 1 method mark for splitting the area into a
rectangle and a semicircle and obtaining at least one of the areas
correctly; 1 mark for finding the total area of the floor; 1 method
mark for attempting to find the volume (given even if '10' used
instead of '0.1', i.e. if the conversion from cm to m is forgotten);
1 mark for the final answer correct to the nearest pound.]

Remember to always split up odd-looking shapes into easy shapes, like
rectangles — it makes it much easier to work out PERIMETERS, AREAS
and VOLUMES.

12 $3x + 4y = 43$ eqn 1
$9x - 6y = 30$ eqn 2

Multiply eqn 1 by 3 to get $9x$ in both equations:
$9x + 12y = 129$ eqn 3
$9x - 6y = 30$ eqn 4

eqn 3 – eqn 4:
$18y = 99$
$y = 5.5$

Sub $y = 5.5$ into eqn 1:
$3x + (4 \times 5.5) = 43$
$3x + 22 = 43$
$3x = 21$
$x = 7$

(Other methods possible.)
[4 marks available — 1 method mark for rearranging the equations to make either the x or the y terms equal OR for rearranging one equation and substituting into the other;
1 mark for eliminating either x or y successfully; 1 mark for finding the value of x; 1 mark for finding the value of y.]

There's a lot going on in SIMULTANEOUS EQUATIONS, so it's always worth substituting your answers back into the equations — just to check you've not gone wrong anywhere (highly unlikely).

13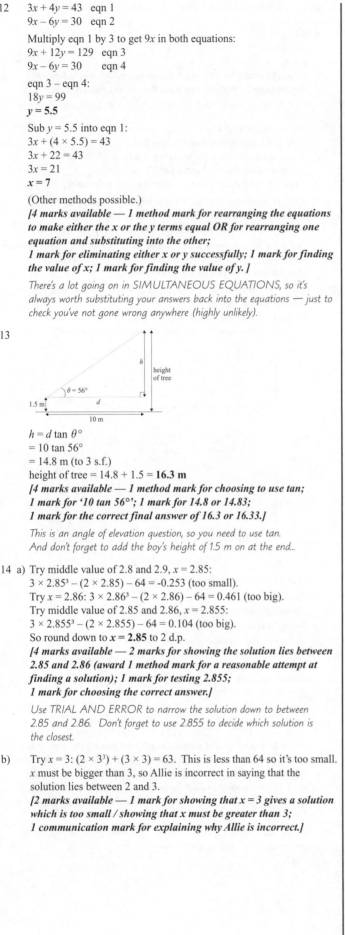

$h = d \tan \theta°$
$= 10 \tan 56°$
$= 14.8$ m (to 3 s.f.)
height of tree $= 14.8 + 1.5 = \textbf{16.3 m}$
[4 marks available — 1 method mark for choosing to use tan; 1 mark for '10 tan 56°'; 1 mark for 14.8 or 14.83; 1 mark for the correct final answer of 16.3 or 16.33.]

This is an angle of elevation question, so you need to use tan. And don't forget to add the boy's height of 1.5 m on at the end…

14 a) Try middle value of 2.8 and 2.9, $x = 2.85$:
$3 \times 2.85^3 - (2 \times 2.85) - 64 = -0.253$ (too small).
Try $x = 2.86$: $3 \times 2.86^3 - (2 \times 2.86) - 64 = 0.461$ (too big).
Try middle value of 2.85 and 2.86, $x = 2.855$:
$3 \times 2.855^3 - (2 \times 2.855) - 64 = 0.104$ (too big).
So round down to $x = \textbf{2.85}$ to 2 d.p.
[4 marks available — 2 marks for showing the solution lies between 2.85 and 2.86 (award 1 method mark for a reasonable attempt at finding a solution); 1 mark for testing 2.855; 1 mark for choosing the correct answer.]

Use TRIAL AND ERROR to narrow the solution down to between 2.85 and 2.86. Don't forget to use 2.855 to decide which solution is the closest.

b) Try $x = 3$: $(2 \times 3^3) + (3 \times 3) = 63$. This is less than 64 so it's too small.
x must be bigger than 3, so Allie is incorrect in saying that the solution lies between 2 and 3.
[2 marks available — 1 mark for showing that x = 3 gives a solution which is too small / showing that x must be greater than 3; 1 communication mark for explaining why Allie is incorrect.]

15 a) Frequency = frequency density × class width

Amount spent on clothes per year (£)	Frequency
$0 \leq x < 50$	$50 \times 0.1 = \textbf{5}$
$50 \leq x < 100$	$50 \times 0.2 = \textbf{10}$
$100 \leq x < 120$	$20 \times 0.7 = \textbf{14}$
$120 \leq x < 140$	$20 \times 1.0 = \textbf{20}$
$140 \leq x < 160$	$20 \times 0.9 = \textbf{18}$
$160 \leq x < 200$	$40 \times 0.3 = \textbf{12}$
$200 \leq x < 300$	$100 \times 0.1 = \textbf{10}$
$x \geq 300$	$\textbf{0}$

[3 marks available — 3 marks for correctly calculating all 8 frequencies. Otherwise 1 mark for showing a calculation involving 'frequency density × class width'; 1 mark for correctly working out all of the class widths.]

b)

Amount spent on clothes per year (£)	Frequency	Mid-interval value	Frequency × mid-interval
$0 \leq x < 50$	5	25	125
$50 \leq x < 100$	10	75	750
$100 \leq x < 120$	14	110	1540
$120 \leq x < 140$	20	130	2600
$140 \leq x < 160$	18	150	2700
$160 \leq x < 200$	12	180	2160
$200 \leq x < 300$	10	250	2500
Totals	89		12 375

$$\text{mean} = \frac{\text{overall total}}{\text{frequency total}} = \frac{12\,375}{89} = \textbf{£139.04}$$

[3 marks available — 1 mark for a correct 'frequency × mid-interval values' column; 1 mark for showing 'mean = 12 375 ÷ 89'; 1 mark for the correct final answer.]

16 Compound interest formula: $N = N_0(1 + \frac{r}{100})^n$
Shop A = $1200 (1.04)^2 = \textbf{£1297.92}$
Shop B = $1200 (1.037)^2 + £100 = £1290.44 + £100$
$= \textbf{£1390.44}$
Faiza and Andrew should choose **Shop A**.
[4 marks available — 1 mark for attempting to use compound interest formula; 1 mark for showing '1200 (1.04)²'; 1 mark for showing '1200 (1.037)²' and '+ 100'; 1 mark for choosing Shop A.]

Don't judge COMPOUND INTEREST by its formula — it's a lot easier to use than it looks.

17 Length x is between 39.5 cm and 40.5 cm.
Volume is between 79 995 cm³ and 80 005 cm³ (since it's correct to 4 s.f., that means it's correct to the nearest 10 cm³).
$y = \text{volume}/x^2$
Largest value $y = \dfrac{\text{biggest volume}}{(\text{smallest length})^2} = \dfrac{80\,005}{39.5^2} = 51.277 = \textbf{51.3 cm}$
Smallest value $y = \dfrac{\text{smallest volume}}{(\text{biggest length})^2} = \dfrac{79\,995}{40.5^2} = 48.770 = \textbf{48.8 cm}$
[5 marks available — 1 mark for stating largest/smallest values of either of the original quantities; 1 mark for either combining biggest with smallest and smallest with biggest or for trying all 4 combinations; 1 mark for correctly finding the smallest possible y value; 1 mark for finding the largest possible y value; 1 mark for both answers given correctly to 3 s.f.]

18 Angle QPC = $90° - 55° = 35°$, as the angle between the radius CP and the tangent is 90°.

Angle PQR is 80°, as angle at the centre is twice the angle at circumference.

Triangle PQC is isosceles, as CP and CQ are both the radius. So angle CQP is 35°, like angle QPC.

Angle CQR = $80° - 35° = 45°$, as angle PQR = angle CQP + angle CQR.

Triangle QRC is isosceles as CR and CQ are both the radius. So angle CRQ is 45°, like angle CQR.

[5 marks available — 1 mark for finding angle QPC; 1 mark for finding angle PQR; 1 mark for attempting to use isosceles triangles for either PQC or QRC; 1 mark for finding CQR; 1 mark for the correct final answer.]

This isn't the only way to do this question, so if you get the right answer by another route, you'll still get the marks. Also, remember, you can always draw extra lines on diagrams to help you — for this question, drawing in line CQ makes it a bit easier.

19 Let the height be x and the length be $x + 4$

So $\frac{1}{2}x(x+4) = 13$

$x^2 + 4x = 26$
$x^2 + 4x - 26 = 0$

This does not factorise, so use the quadratic equation formula:
$a = 1, b = 4, c = -26$

$x = \frac{-b \pm \sqrt{b^2 - 4ac}}{2a}$

$x = \frac{-4 \pm \sqrt{4^2 - (4 \times 1 \times -26)}}{2 \times 1}$

$x = \frac{-4 \pm \sqrt{16 - (-104)}}{2}$

$x = \frac{-4 \pm \sqrt{120}}{2} = 3.48$ or -7.48

But x cannot be negative as it is the height of a triangle. So $x = 3.48$ cm.

[5 marks available — 1 mark for getting '1/2x(x + 4) = 13' (or for '1/2x(x – 4) = 13'); 1 mark for multiplying out and rearranging to get 0 on one side; 1 mark for a fully correct substitution into the quadratic formula; 1 mark for getting either of the two solutions; 1 mark for indicating the correct solution is 3.48 cm only.]

If you're asked to find the solution to a quadratic and give your answer to a particular number of significant figures, be prepared to use the quadratic formula — if the quadratic factorised, it would give you an exact number... sneaky but true.

20 Gift box A:
$x^2 = 12^2 + 11^2$
$x^2 = 144 + 121 = 265$
$y^2 = x^2 + 12^2$
$y^2 = 265 + 144 = 409$, so $y = 20.223...$

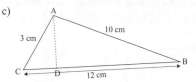

Gift box B: $x^2 = 13^2 + 11^2$
$x^2 = 169 + 121 = 290$
$y^2 = x^2 + 10^2$
$y^2 = 290 + 100 = 390$, so $y = 19.748...$

Gift box A has a diagonal length of over 20 cm so it is big enough to fit the sweets in and it is cheaper than gift box C. So Zahra should buy gift box **A**.

[5 marks available — 1 method mark for using Pythagoras theorem and writing a valid equation; 1 mark for finding the first diagonal length 'x' for at least two of the gift boxes (x^2 is also acceptable); 1 mark for finding 'y²' for at least two of the gift boxes; 1 mark for calculating the correct diagonal length 'y' for at least two of the gift boxes; 1 mark for the choosing the correct gift box.]

Once you've spotted it's 3D PYTHAGORAS THEOREM, this question's not too bad.

21 a) Use the cosine rule, as the triangle is not right-angled and we don't have any angles. Cosine rule: $a^2 = b^2 + c^2 - 2bc \cos A$.
So $12^2 = 3^2 + 10^2 - 2 \times 3 \times 10 \cos A$
$144 = 9 + 100 - 60\cos A$
$60\cos A = 109 - 144 = -35$
$\cos A = -35/60$
$A = 125.7°$

[3 marks available — 1 method mark for using the cosine rule and substituting into it; 1 mark for finding the value of cos A (mark may also be awarded if only the final answer is given); 1 mark for the correct final answer.]

b) Area of triangle = ½ bc sinA
= ½ × 3 × 10 × sin 125.7°
= 12.18... = **12 cm²**

[2 marks available — 1 method mark for use of '½ bc sin A'; 1 mark for the correct final answer (give follow-through marks for use of sine of angle found in part a).]

c)

Area of triangle = ½ BC × AD
So ½ × 12 × AD = area from part b).
AD = **2 cm**

[2 marks available — 1 mark for correctly labelling 'D'; 1 mark for the correct answer.]

Set 2 Paper 1 — Non-calculator

1 a) $\frac{11}{17} + \frac{25}{17} = \frac{36}{17} = 2\frac{2}{17}$

[2 marks available — 1 mark for writing $\frac{36}{17}$; 1 mark for correct final answer.]

b) i) $2\frac{4}{5} = \frac{14}{5} = \frac{126}{45}$

$\frac{7}{9} = \frac{35}{45}$

$\frac{126}{45} - \frac{35}{45} = \frac{91}{45}$ or $2\frac{1}{45}$

[2 marks available — 1 mark for writing $2\frac{4}{5}$ as '126/45' or for writing '7/9 as 35/45'; 1 mark for correct final answer given as either 91/45 or $2\frac{1}{45}$.]

ii) $4\frac{11}{12} = \frac{59}{12}$

$\frac{5}{6} \div \frac{59}{12} = \frac{5}{6} \times \frac{12}{59} = \frac{5}{\cancel{6}} \times \frac{\cancel{12}^2}{59} = \frac{10}{59}$

[2 marks available — 1 mark for showing '5/6 × 12/59'; 1 mark for correct final answer.]

Ahh — a good old FRACTIONS question — wherever you go, they're always there.

2 a) $10 - 3x > 5x - 6$
$10 > 8x - 6$
$16 > 8x$
$2 > x$ or $x < 2$
[1 mark]

b)

[1 mark]

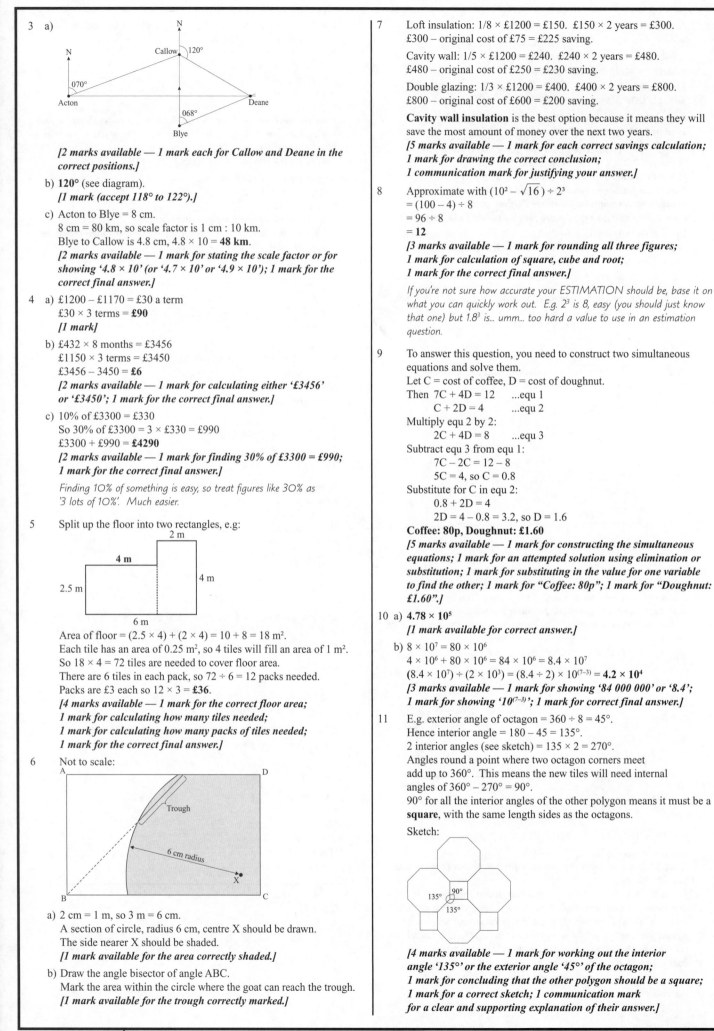

3 a)

[2 marks available — 1 mark each for Callow and Deane in the correct positions.]

b) **120°** (see diagram).
[1 mark (accept 118° to 122°).]

c) Acton to Blye = 8 cm.
8 cm = 80 km, so scale factor is 1 cm : 10 km.
Blye to Callow is 4.8 cm, 4.8 × 10 = **48 km**.
[2 marks available — 1 mark for stating the scale factor or for showing '4.8 × 10' (or '4.7 × 10' or '4.9 × 10'); 1 mark for the correct final answer.]

4 a) £1200 – £1170 = £30 a term
£30 × 3 terms = **£90**
[1 mark]

b) £432 × 8 months = £3456
£1150 × 3 terms = £3450
£3456 – 3450 = **£6**
[2 marks available — 1 mark for calculating either '£3456' or '£3450'; 1 mark for the correct final answer.]

c) 10% of £3300 = £330
So 30% of £3300 = 3 × £330 = £990
£3300 + £990 = **£4290**
[2 marks available — 1 mark for finding 30% of £3300 = £990; 1 mark for the correct final answer.]

Finding 10% of something is easy, so treat figures like 30% as '3 lots of 10%'. Much easier.

5 Split up the floor into two rectangles, e.g:

Area of floor = (2.5 × 4) + (2 × 4) = 10 + 8 = 18 m².
Each tile has an area of 0.25 m², so 4 tiles will fill an area of 1 m².
So 18 × 4 = 72 tiles are needed to cover floor area.
There are 6 tiles in each pack, so 72 ÷ 6 = 12 packs needed.
Packs are £3 each so 12 × 3 = **£36**.
[4 marks available — 1 mark for the correct floor area; 1 mark for calculating how many tiles needed; 1 mark for calculating how many packs of tiles needed; 1 mark for the correct final answer.]

6 Not to scale:

a) 2 cm = 1 m, so 3 m = 6 cm.
A section of circle, radius 6 cm, centre X should be drawn.
The side nearer X should be shaded.
[1 mark available for the area correctly shaded.]

b) Draw the angle bisector of angle ABC.
Mark the area within the circle where the goat can reach the trough.
[1 mark available for the trough correctly marked.]

7 Loft insulation: 1/8 × £1200 = £150. £150 × 2 years = £300.
£300 – original cost of £75 = £225 saving.

Cavity wall: 1/5 × £1200 = £240. £240 × 2 years = £480.
£480 – original cost of £250 = £230 saving.

Double glazing: 1/3 × £1200 = £400. £400 × 2 years = £800.
£800 – original cost of £600 = £200 saving.

Cavity wall insulation is the best option because it means they will save the most amount of money over the next two years.
[5 marks available — 1 mark for each correct savings calculation; 1 mark for drawing the correct conclusion; 1 communication mark for justifying your answer.]

8 Approximate with $(10^2 - \sqrt{16}) \div 2^3$
$= (100 - 4) \div 8$
$= 96 \div 8$
$= \mathbf{12}$
[3 marks available — 1 mark for rounding all three figures; 1 mark for calculation of square, cube and root; 1 mark for the correct final answer.]

If you're not sure how accurate your ESTIMATION should be, base it on what you can quickly work out. E.g. 2^3 is 8, easy (you should just know that one) but 1.8^3 is... umm... too hard a value to use in an estimation question.

9 To answer this question, you need to construct two simultaneous equations and solve them.
Let C = cost of coffee, D = cost of doughnut.
Then 7C + 4D = 12   ...equ 1
     C + 2D = 4   ...equ 2
Multiply equ 2 by 2:
     2C + 4D = 8   ...equ 3
Subtract equ 3 from equ 1:
     7C – 2C = 12 – 8
     5C = 4, so C = 0.8
Substitute for C in equ 2:
     0.8 + 2D = 4
     2D = 4 – 0.8 = 3.2, so D = 1.6
Coffee: 80p, Doughnut: £1.60
[5 marks available — 1 mark for constructing the simultaneous equations; 1 mark for an attempted solution using elimination or substitution; 1 mark for substituting in the value for one variable to find the other; 1 mark for "Coffee: 80p"; 1 mark for "Doughnut: £1.60".]

10 a) **4.78×10^5**
[1 mark available for correct answer.]

b) $8 \times 10^7 = 80 \times 10^6$
$4 \times 10^6 + 80 \times 10^6 = 84 \times 10^6 = 8.4 \times 10^7$
$(8.4 \times 10^7) \div (2 \times 10^3) = (8.4 \div 2) \times 10^{(7-3)} = \mathbf{4.2 \times 10^4}$
[3 marks available — 1 mark for showing '84 000 000' or '8.4'; 1 mark for showing '$10^{(7-3)}$'; 1 mark for correct final answer.]

11 E.g. exterior angle of octagon = 360 ÷ 8 = 45°.
Hence interior angle = 180 – 45 = 135°.
2 interior angles (see sketch) = 135 × 2 = 270°.
Angles round a point where two octagon corners meet add up to 360°. This means the new tiles will need internal angles of 360° – 270° = 90°.
90° for all the interior angles of the other polygon means it must be a **square**, with the same length sides as the octagons.
Sketch:

[4 marks available — 1 mark for working out the interior angle '135°' or the exterior angle '45°' of the octagon; 1 mark for concluding that the other polygon should be a square; 1 mark for a correct sketch; 1 communication mark for a clear and supporting explanation of their answer.]

12 $$\frac{3 \times 4(x+3)}{2 \times 3} - \frac{2(x+2)}{3 \times 2} = 3$$

$$\frac{12(x+3)}{6} - \frac{2(x+2)}{6} = 3$$

$12(x+3) - 2(x+2) = 3 \times 6$
$12x + 36 - 2x - 4 = 18$
$10x + 32 = 18$
$10x = -14$
$x = -1.4$
[4 marks available — 1 method mark for using the correct equivalent fractions; 1 method mark for multiplying by 6; 1 mark for correct expansion and simplification of the brackets; 1 mark for the correct final answer.]

Watch out this is an EQUIVALENT FRACTIONS question.

13 The median is 18 without the system and 14 with the new system. The interquartile range is 14 without the system and 10 with the new system. (The range is 30 without and 28 with.)
Comparing the two systems, the median shows that the phone was answered more quickly with the new system in place.
The interquartile range for the new system was smaller than for the old system, showing that the time taken to answer calls was more consistent with the new system.
In conclusion, the box plots suggest that *Parkers Patios* should keep the new phone system because it is more efficient than the old system and calls will get answered more quickly.
[6 marks available — 1 mark for each median; 1 mark for each interquartile range or range; 1 mark for saying they should keep the new phone system; 1 communication mark for justifying your answer.]

14 a) There are 12 sweets altogether. 4 are red and 3 are yellow.
P(picking a red OR yellow sweet) = 4/12 + 3/12 = **7/12**.
[1 mark]

b) Replacing the sweet means the probability stays the same for the second pick.
P(picking a green sweet AND picking a green sweet)
= 5/12 × 5/12 = **25/144**.
[2 marks available — 1 mark for '5/12 × 5/12'; 1 mark for correct answer.]

c) Not replacing the sweet means the probability changes for the second pick.
Find the probability of picking 2 sweets the <u>same</u> colour, then subtract this probability from 1.
P(red, red) = 4/12 × 3/11 = 12/132
P(green, green) = 5/12 × 4/11 = 20/132
P(yellow, yellow) = 3/12 × 2/11 = 6/132
P(2 different coloured sweets) = 1 − P(2 sweets the same colour)
= 1 − (12/132 + 20/132 + 6/132)
= 1 − 38/132
= **94/132** or **47/66**
[4 marks available — 1 mark for calculating one probability correctly, '12/132', '20/132' or '6/132'; 1 mark for calculating all three probabilities correctly; 1 mark for showing '1 − 38/132'; 1 mark for the correct final answer.]

Learn when to use the '1 − P' rule... it's a lot quicker than finding all the different combinations and adding them.

15 a)

x	-3	-2	-1	0	1	2	3
$y = 3x^2 - 2$	25	**10**	1	**-2**	1	**10**	25

[2 marks available — 2 marks for all 3 correct answers; lose 1 mark per mistake made.]

b)
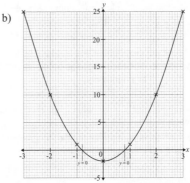

[2 marks available — 1 mark for correctly plotting all points for the graph 'y = 3x² – 2'; 1 mark for a smooth curve connecting all the points.]

c) $3x^2 - 2 = 0$, so $y = 0$.
From the graph, $x = -0.8$ and $x = 0.8$.
[1 mark for both correct answers ± 0.1]

16 a) x-coordinate: $(3 + 8) \div 2 = 5.5$
y-coordinate: $(9 + 24) \div 2 = 16.5$
The coordinates of the mid-point are **(5.5, 16.5)**.
[2 marks available — 1 mark for the correct x-coordinate; 1 mark for the correct y-coordinate.]

b) Gradient = $\frac{24-9}{8-3} = \frac{15}{5} = 3$

The gradient of the line is **3**.
[1 mark]

c) Equation of line AB: $y = mx + c$.
From part b): m (gradient) = 3, so $y = 3x + c$.
A (3, 9) and B (8, 24) so $y = 9$ when $x = 3$.
$y = 3x + c$
$9 = 3 \times 3 + c$
$9 = 9 + c$
So c (y-intercept) = 0
$y = 3x + 0$ so the equation of line AB is $y = 3x$.
[2 marks available — 1 mark for 'y = mx + c' or implied by substitution of the gradient; 1 mark for correct final answer.]

d) i) Parallel lines have the same gradient (m), so gradient = **3**.
[1 mark]

ii) Gradient of perpendicular line = $\frac{-1}{m} = \frac{-1}{3} = $ **-1/3**
[1 mark]

17 a) Angle ABC = angle EDC (alternate angles between parallel lines).
Angle BAC = angle DEC (alternate angles between parallel lines).
Angle ACB = angle ECD (vertically opposite angles).
Hence triangles ABC and EDC are similar (AAA rule).
[3 marks available — 1 mark for 2 pairs of equal angles, with a reason for at least one pair; 1 mark for statement of similarity with a reason; 1 communication mark for a clear explanation.]

b) You only know for certain that the triangles have 1 equal length side (BE). To show congruence, you need to show that the two triangles are the same size and the same shape. There is not enough information given in the diagram because you need to prove that either 2 sides and an angle, or 2 angles and a side, or 3 sides are equal to show congruence.
[2 marks available — 1 mark for explaining congruence or showing an understanding of congruence; 1 mark for stating that there is not enough information here to show congruence.]

18 a)

Age (a) (years)	$0 \le a < 20$	$20 \le a < 30$	$30 \le a < 35$	$35 \le a < 40$	$40 \le a < 55$	$55 \le a < 80$
Frequency	20	30	40	30	45	25
Class width	20	**10**	5	5	15	25
Frequency density	1	**3**	**8**	6	3	1

[2 marks available — 2 marks for table correctly filled in; lose 1 mark for each error.]

b) Median = $(190 + 1) \div 2 = 95.5$. The class containing the age of the 95.5th (middle) member is $\mathbf{35 \leqslant a < 40}$.
[1 mark]

c) i) Draw a line at 60, 4/5 of the members in this class are over 60 = 4/5 of 25 = **20 members**. OR each large square represents 10 members, so two squares = **20 members**.
[1 mark]

ii) The data has been grouped, so all the members in the class could be over 60 or under 60 because the group goes from 55 up to 80.
[2 marks available — 1 mark for stating the data is grouped; 1 communication mark for a correct explanation.]

If you've not learnt it by now… those examiners love testing you with a good old HISTOGRAM question.

19 a)
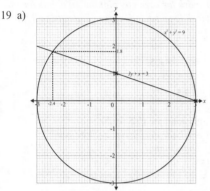

The graph is a circle, centre (0,0) and radius 3 units.
The quickest way to draw it is to use a pair of compasses.
[2 marks available — 1 mark for drawing a circle with the correct centre; 1 mark for drawing a circle with the correct radius.]

This sneaky circle graph is hidden away with the HARDER GRAPHS TO LEARN — so I hope you learnt it.

b) Rearrange the equation in the form of $y = mx + c$:

$3y + x = 3, y + \dfrac{x}{3} = 1, y = \dfrac{-x}{3} + 1$

So y-intercept (c) = **1** and the gradient (m) is **-1/3**.
The graph should cut the y-axis at 1 and the x-axis at 3.
[2 marks available — 1 mark for correctly rearranging the equation; 1 mark for a correctly drawn graph (1 follow-through mark if the graph is drawn correctly but from an incorrect rearrangement, e.g. 'y = 1 + 1/3 x' or 'y = 3 – 1/3x').]

c) The coordinates are **(3, 0)** and **(-2.4, 1.8)**.
[2 marks available — 1 mark for each correct pair of coordinates. Give 1 mark if both pairs of coordinates are correctly recorded from an incorrectly drawn line.]

20 a) Factorise the denominator first, as this will give you one of the brackets for the numerator.
Denominator: $x^2 - 4 = (x - 2)(x + 2)$
$10 \div 2 = 5$,
so $x^2 + 3x - 10 = (x - 2)(x + 5)$.

Cancel the common factor of $(x - 2)$,
so the answer is $\dfrac{x + 5}{x + 2}$

[3 marks available — 1 mark for correctly factorising the denominator; 1 mark for correctly factorising the numerator; 1 mark for cancelling down to get the correct final answer.]

b) $x + 5 = 2(x + 2)$
$x + 5 = 2x + 4$
$\mathbf{x = 1}$

Alternatively, $x^2 + 3x - 10 = 2(x^2 - 4)$
$x^2 + 3x - 10 = 2x^2 - 8$
$x^2 - 3x + 2 = 0$
$x = 2$ or $x = 1$
But if you substitute 2 back into the original expression, the denominator is $4 - 4 = 0$, so 2 cannot be a solution. So $x = 1$.
[2 marks available — 1 mark for using either of the methods shown above; 1 mark for the correct final answer.]

21 Length of arc = $\dfrac{\text{angle}}{360} \times$ circumference of full circle (πd)

$= \dfrac{240}{360} \times 24\pi = \dfrac{24}{36} \times 24\pi = \dfrac{2}{3} \times \dfrac{24}{1}\pi = \dfrac{48}{3}\pi = 16\pi$ cm

The circumference of the top of the cone is 16π cm.
$2\pi r = 16\pi$ cm,
$\mathbf{r = 8}$ cm.
[4 marks available — 1 mark for stating the formula for the length of arc or implied by substitution; 1 mark for correct substitution and simplification, leaving in terms of π; 1 mark for stating or implying that this is the circumference of the top of the cone; 1 mark for finding the correct value of r.]

This is a non-calculator paper, so leave π in your calculations.

Set 2 Paper 2 — Calculator

1 a) The ratio of the amounts the girls get is 36:63
This cancels down to 4:7 (by dividing by 9).
The ratio of the ages is 8: ?
4:7 is the same ratio as 8:14
Sal is **14**.
[2 marks available — 1 method mark for attempting to use equivalent ratios; 1 mark for the correct answer.]

b) Dividing in the ratio 4:3 means there are 7 parts altogether.
7 parts = £63, so 1 part = £9.
The two parts are split into $4 \times$ £9 and $3 \times$ £9.
So she gives her boyfriend **£27**.
[2 marks available — 1 mark for the idea of 7 parts and trying to find 1 part (this can also be awarded if £9 is seen); 1 mark for the correct final answer.]

2 a) $4x + 8 = 16$
$4x = 8$
$\mathbf{x = 2}$
[1 mark for the correct answer.]

b) $3x - 4 = 6x + 8$
$3x - 6x = 8 + 4$
$-3x = 12$
$x = \dfrac{12}{-3}$
$\mathbf{x = -4}$
[2 marks available — 1 method mark for '-3x = 12'; 1 mark for the correct final answer.]

3 a) 1 kg = 2.2 lb, so 1.1 lb = **0.5 kg** *[1 mark]*

b) Yes, 1 litre = 1.75 pints so she has enough milk to make the bake.
[1 mark for stating 'yes' with a correct explanation.]

You need to know your METRIC AND IMPERIAL CONVERSIONS for this question.

c) i) $F = \dfrac{9C}{5} + 32$
$F - 32 = \dfrac{9C}{5}$
$5(F - 32) = 9C$

$\mathbf{C = \dfrac{5}{9}(F - 32)}$

[1 mark for the correct final answer.]

ii) $C = \dfrac{5}{9}(F - 32)$; F = 390

$C = \dfrac{5}{9}(390 - 32)$

$C = \dfrac{5 \times 358}{9}$

$C = 198.88...$ so Mary should set her oven to **200 °C**.
[2 marks available — 1 method mark for the correct substitution; 1 mark for the correct answer. Award both marks if the answer is correct but no method is shown.]

4 a) Total number of customers = 200. 360° ÷ 200 = 1.8, so multiply the number of customers by 1.8 to get each angle in the pie chart:

Response	No. customers	**Angle in pie chart**
Agree strongly	70	**126** (70 × 1.8)
Mostly agree	80	**144** (80 × 1.8)
Neither agree nor disagree	25	**45** (25 × 1.8)
Mostly disagree	20	**36** (20 × 1.8)
Strongly disagree	5	**9** (5 × 1.8)
Total	**200**	**360**

[4 marks available — 1 method mark for using the correct method of finding angles; 1 mark for all angles correct;
2 marks for a correct pie chart (lose 1 mark for each angle wrongly drawn; lose 1 mark if no labels are drawn).]

b) E.g. "No it's not reasonable — you can't tell from the pie charts how many responses are shown by the Asco pie chart, so the pie charts cannot be compared."
Or, e.g. "Yes it's reasonable — around 75% of people mostly or strongly agreed with the statement for Rollinsons, but only approximately 55% of people mostly or strongly agreed with the statement for Asco."
[2 marks available — 1 mark for stating "Yes" or "No" with a reason; 1 communication mark for clearly justifying your answer.]

5 a) Translation by the vector $\begin{pmatrix} -10 \\ -9 \end{pmatrix}$
OR translation by 10 units to the left/in the negative x direction and by 9 units down/in the negative y direction.
[3 marks available — 1 mark for stating 'translation';
1 mark for stating '10 units left' or '-10 units horizontally'
or writing the correct component of the vector;
1 mark for stating '9 units down' or '-9 units vertically'
or writing the correct component of the vector.]

b)
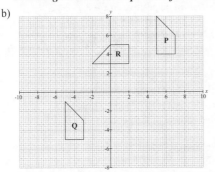

[2 marks available — 1 mark for drawing shape R rotated 90° anti-clockwise; 1 mark for drawing at least two points of shape R in the correct location, indicating the correct centre of rotation has been used OR for construction lines joining the shape to the centre of rotation; full marks for completely correct coordinates of shape R.]

You're allowed tracing paper in the exam, so use it — it's an easy way to do rotation questions. Just trace the shape, put your pencil point on the point you're rotating about and then move the tracing paper, making sure the new points you draw are on exact coordinates.

6 a) $\dfrac{14.82983871...}{4.397496} = $ **3.372337054**

(Your calculator may give an answer to a different number of d.p.)
[2 marks available — 2 marks for the correct answer; otherwise 1 mark for calculating the numerator or denominator correctly.]

Remember BODMAS and remember not to round during a calculation.

b) **3.37**
[1 mark]

7 3 drawer chest   £48
large tallboy   + £140
bedside table   + £28
desk   + £50
bookcase   + £43
pine bed   + £140
total   £449
add VAT = 20% of £449 = £449 × 0.2 = £89.80
total including VAT = £449 + £89.80 = £538.80
less 20% discount = £538.80 × 0.8 = **£431.04**
[4 marks available — 1 mark for correctly adding the selection of prices from the table; 1 method mark for calculating VAT;
1 method mark for calculating the discount; 1 mark for the correct final answer.]

Don't try to do all this in one go — there are 4 marks up for grabs here so do the calculation in steps, show your working and double check your final answer.

8
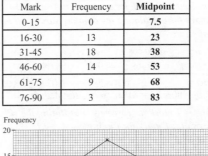
BC = **6.5 cm**

[3 marks available — 2 marks for accurately drawing the triangle (1 mark if 1 measurement is incorrect, 0 marks if 2 or more sides are drawn incorrectly. Allow an error of 1 mm in length and 1° in angle); 1 mark for the correct length of BC (allow 1 mm either way).]

9 Find the midpoint of each interval and plot the midpoint against the frequency. Midpoints should be plotted at 7.5, 23, 38, 53, 68 and 83.

Mark	Frequency	**Midpoint**
0-15	0	**7.5**
16-30	13	**23**
31-45	18	**38**
46-60	14	**53**
61-75	9	**68**
76-90	3	**83**

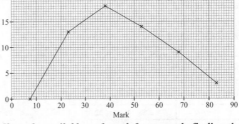

[3 marks available — 1 mark for correctly finding the midpoints;
1 mark for correctly plotting the midpoints against the frequency;
1 mark for correctly joining the points to make a polygon. Lose 1 mark for mistakes in finding the midpoints if the rest of the graph is drawn correctly.]

10 a) $V = \pi r^2 h$
$V = \pi \times 10^2 \times 10$
$V = \mathbf{1000\pi}$

[2 marks available — 1 mark for the formula 'V = π r²h';
1 mark for the correct final answer.]

b) density = mass ÷ volume
mass = 3 kg = 3000 g
volume = 1000π = 3141.59... cm³
density = 3000 ÷ 3141.59... = 0.9549... = **0.95 g/cm³**
[3 marks available — 1 mark for showing that 'density = mass ÷ volume'; 1 mark for both '3000 g' and '3141.59... cm³' (or 1000π); 1 mark for correct final answer.]

c) Total area = top surface area + curved surface area = $\pi r^2 + 2\pi rh$
top surface area = $\pi r^2 = \pi \times 10^2 = 100\pi$ cm^2
curved surface area = $2\pi rh = 2 \times \pi \times 10 \times 10 = 200\pi$ cm^2
total area = $100\pi + 200\pi = 300\pi = 942.477...$ cm^2.
Packs required = $942.477... \div 200 \approx$ **5 packs of marzipan**
[4 marks available — 1 mark for correctly calculating '$\pi r^2 = 100\pi$'
(or 314.15...'); 1 mark for correctly calculating '$2\pi rh = 200\pi$' (or
628.31...'); 1 mark for finding the total area; 1 mark for the correct
final answer.]

11 a) $3x(xy - 3y^3 + 2x^2)$
[2 marks available for correct answer.
Give 1 mark for '$3(x^2y - 3xy^3 + 2x^3)$' or '$x(3xy - 9y^3 + 6x^2)$'.]

b) $(5x - 4)(x + 2)$
[2 marks available for correct answer.
Give 1 mark for showing double brackets '$(5x\ \ \ )(x\ \ \ )$'.]

12 Total calories Ken needs to use/not eat = $6 \times 3500 = 21\,000$ kCal.
Amount saved by not eating snacks for 4 weeks
= $1800 + 3200 + 6100 + 1450 = 12\,550$ kCal.
Total to burn through exercise = $21\,000 - 12\,550 = 8450$ kCal.
Total to burn per week = $8450 \div 4 = 2112.5$ kCal
Total to burn per exercise session = $2112.5 \div 3 = 704$ (to 3 s.f.)
So he should join the **spin class**.

[4 marks available — 1 mark for finding the total number of
calories to be used/not eaten; 1 mark for finding the total kCal he
needs to use up through exercise; 1 mark for finding the total kCal
he needs to be use up per exercise class; 1 mark for stating Ken
should do the spin class with supporting calculations.]

13 Account 1: £5000 × 1.035^2 = £5356.13
Account 2: £5000 − £150 = £4850, £4850 × 1.05^2 = £5347.13
Since £5356.13 > £5347.13, the better option is **Account 1**.
Sarah will make 5356.13 − 5347.13 = **£9.00**.

[4 marks available — 1 mark for correctly working out the value of
investing in Account 1; 1 mark for correctly working out the value
of investing in Account 2; 1 mark for stating that Account 1 is the
better option; 1 mark for calculating the amount Sarah will make by
choosing Account 1.]

14 a) Cosine Rule: $a^2 = b^2 + c^2 - 2bc\,\cos A$
$$\cos A = \frac{b^2 + c^2 - a^2}{2bc} = \frac{6^2 + 8.5^2 - 9.6^2}{2 \times 6 \times 8.5} = \frac{16.09}{102} = 0.1577...$$
$A = \cos^{-1}(0.1577...) = 80.92... = $ **81°** to the nearest degree.

[3 marks available — 1 mark for correctly substituting the values
into the cosine formula; 1 mark for finding $\cos A = 0.1577...$; 1 mark
for the correct answer.]

b) Area of triangle = $\frac{1}{2}cb\sin A$
= $\frac{1}{2} \times 8.5 \times 6 \times \sin 80.92... = 25.18... = $ **25 cm^2**

[2 marks available — 1 mark for correctly substituting values into
the area formula; 1 mark for the correct final answer.]

15 a)

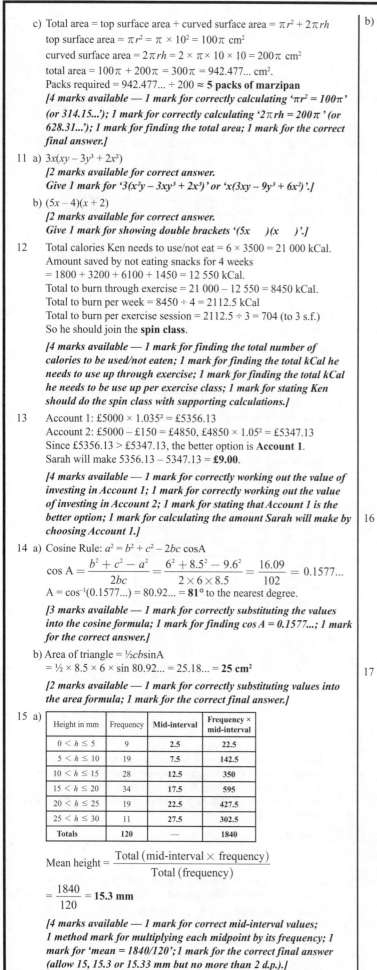

Height in mm	Frequency	Mid-interval	Frequency × mid-interval
$0 < h \leq 5$	9	2.5	22.5
$5 < h \leq 10$	19	7.5	142.5
$10 < h \leq 15$	28	12.5	350
$15 < h \leq 20$	34	17.5	595
$20 < h \leq 25$	19	22.5	427.5
$25 < h \leq 30$	11	27.5	302.5
Totals	120	—	1840

$$\text{Mean height} = \frac{\text{Total (mid-interval} \times \text{frequency)}}{\text{Total (frequency)}}$$

$$= \frac{1840}{120} = \textbf{15.3 mm}$$

[4 marks available — 1 mark for correct mid-interval values;
1 method mark for multiplying each midpoint by its frequency; 1
mark for 'mean = 1840/120'; 1 mark for the correct final answer
(allow 15, 15.3 or 15.33 mm but no more than 2 d.p.).]

b)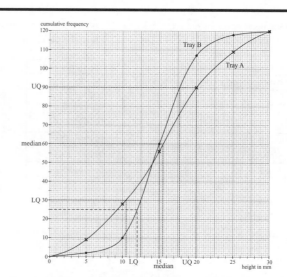

Tray A: median = **15.5 mm**, interquartile range = $20 - 10.5 = $ **9.5 mm**
Tray B: median = **15 mm**
    interquartile range = $17.75 - 12.5 = $ **5.25 mm**
The median for both trays is similar.
Tray A: the cumulative frequency graph is much straighter, so the
height readings are more evenly spread. This is also shown by the
large interquartile range.
Tray B: there is a tight distribution around the median and a much
smaller interquartile range / less variation in growth. The cumulative
frequency graph shows far fewer very high and low readings, so the
heights are less evenly spread.
[6 marks available — 1 mark for each correct median;
1 mark for each correct interquartile range;
2 communication marks for comparing the data.]

c) 25 seedlings are 12 mm and below (see graph),
so $120 - 25 = $ **95 seedlings**.
[1 mark]

16 $48 = 2x \times (x - 5)$
$48 = 2x^2 - 10x$
$24 = x^2 - 5x$
$x^2 - 5x - 24 = 0$
$(x - 8)(x + 3) = 0$
$x = 8$ or $x = -3$
Since $2x$ is a length, x must be positive so $x = 8$.
[5 marks available — 1 mark for '$48 = 2x \times (x - 5)$';
1 mark for '$2x^2 - 10x = 48$'; 1 mark for correct factorisation;
1 mark for the correct solutions; 1 mark for choosing the positive
solution.]

17 Area is between 1.95 m^2 and 2.05 m^2.
Depth is between 19 cm and 21 cm.
Convert depth to m: between 0.19 m and 0.21 m.
The smallest possible volume of the pool is 1.95×0.19
= 0.3705 m^3.
The bucket holds between 9.75 and 10.25 litres.
$35 \times 10.25 = 358.75$ litres = 0.35875 m^3.
No you cannot be certain. If the pool is the smallest possible size and
the bucket is the largest possible size, 35 buckets of water will not
completely fill the pool.
[5 marks available — 2 marks for correct bounds of area, depth and
bucket capacity (award 1 mark if 2 out of 3 are correct);
1 mark for finding the smallest possible volume of the pool;
1 mark for converting litres to m^3 (or m^3 to litres) correctly;
1 communication mark for 'no' with supporting calculations.]

Remember the rule for BOUNDS: +/− half the 'nearest' value.

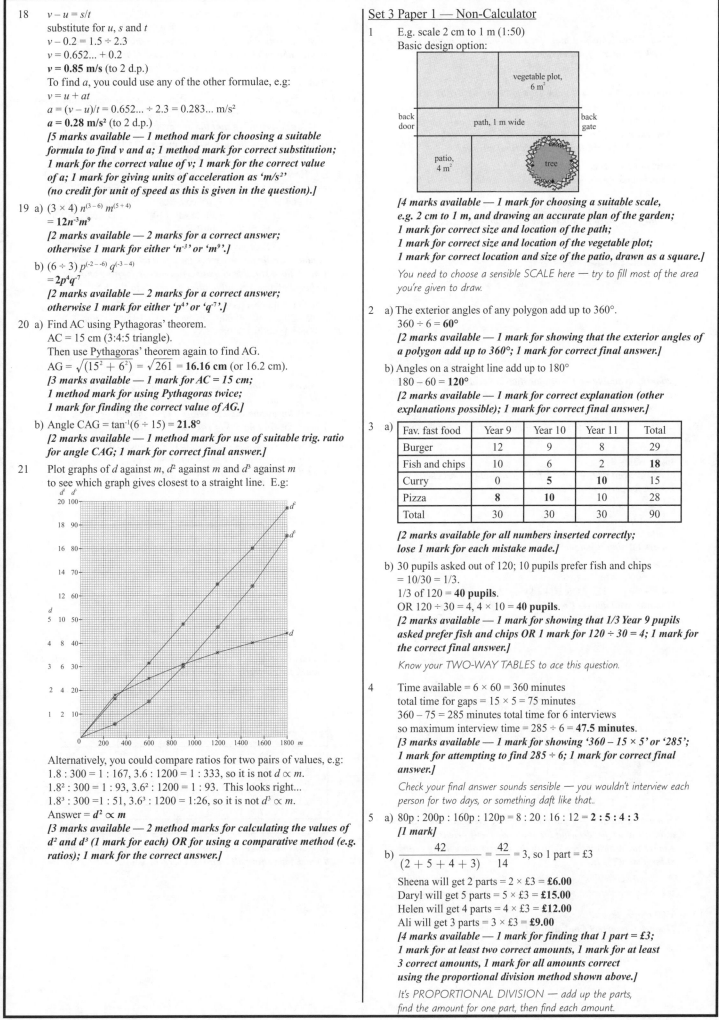

18 $v - u = s/t$

substitute for u, s and t

$v - 0.2 = 1.5 \div 2.3$

$v = 0.652... + 0.2$

$v = 0.85$ m/s (to 2 d.p.)

To find a, you could use any of the other formulae, e.g:

$v = u + at$

$a = (v - u)/t = 0.652... \div 2.3 = 0.283...$ m/s²

$a = 0.28$ m/s² (to 2 d.p.)

[5 marks available — 1 method mark for choosing a suitable formula to find v and a; 1 method mark for correct substitution; 1 mark for the correct value of v; 1 mark for the correct value of a; 1 mark for giving units of acceleration as 'm/s²' (no credit for unit of speed as this is given in the question).]

19 a) $(3 \times 4)\, n^{(3 - 6)}\, m^{(5 + 4)}$

$= \mathbf{12n^{-3}m^9}$

[2 marks available — 2 marks for a correct answer; otherwise 1 mark for either 'n⁻³' or 'm⁹'.]

b) $(6 \div 3)\, p^{(-2 - -6)}\, q^{(-3 - 4)}$

$= \mathbf{2p^4 q^{-7}}$

[2 marks available — 2 marks for a correct answer; otherwise 1 mark for either 'p⁴' or 'q⁻⁷'.]

20 a) Find AC using Pythagoras' theorem.

AC = 15 cm (3:4:5 triangle).

Then use Pythagoras' theorem again to find AG.

$AG = \sqrt{(15^2 + 6^2)} = \sqrt{261} = \mathbf{16.16\ cm}$ (or 16.2 cm).

[3 marks available — 1 mark for AC = 15 cm; 1 method mark for using Pythagoras twice; 1 mark for finding the correct value of AG.]

b) Angle CAG = $\tan^{-1}(6 \div 15) = \mathbf{21.8°}$

[2 marks available — 1 method mark for use of suitable trig. ratio for angle CAG; 1 mark for correct final answer.]

21 Plot graphs of d against m, d^2 against m and d^3 against m to see which graph gives closest to a straight line. E.g:

Alternatively, you could compare ratios for two pairs of values, e.g:

1.8 : 300 = 1 : 167, 3.6 : 1200 = 1 : 333, so it is not $d \propto m$.

1.8² : 300 = 1 : 93, 3.6² : 1200 = 1 : 93. This looks right...

1.8³ : 300 = 1 : 51, 3.6³ : 1200 = 1:26, so it is not $d^3 \propto m$.

Answer = $\mathbf{d^2 \propto m}$

[3 marks available — 2 method marks for calculating the values of d² and d³ (1 mark for each) OR for using a comparative method (e.g. ratios); 1 mark for the correct answer.]

Set 3 Paper 1 — Non-Calculator

1 E.g. scale 2 cm to 1 m (1:50)

Basic design option:

[4 marks available — 1 mark for choosing a suitable scale, e.g. 2 cm to 1 m, and drawing an accurate plan of the garden; 1 mark for correct size and location of the path; 1 mark for correct size and location of the vegetable plot; 1 mark for correct location and size of the patio, drawn as a square.]

You need to choose a sensible SCALE here — try to fill most of the area you're given to draw.

2 a) The exterior angles of any polygon add up to 360°.

$360 \div 6 = \mathbf{60°}$

[2 marks available — 1 mark for showing that the exterior angles of a polygon add up to 360°; 1 mark for correct final answer.]

b) Angles on a straight line add up to 180°

$180 - 60 = \mathbf{120°}$

[2 marks available — 1 mark for correct explanation (other explanations possible); 1 mark for correct final answer.]

3 a)

Fav. fast food	Year 9	Year 10	Year 11	Total
Burger	12	9	8	29
Fish and chips	10	6	2	**18**
Curry	0	**5**	**10**	15
Pizza	**8**	**10**	10	28
Total	30	30	30	90

[2 marks available for all numbers inserted correctly; lose 1 mark for each mistake made.]

b) 30 pupils asked out of 120; 10 pupils prefer fish and chips

= 10/30 = 1/3.

1/3 of 120 = **40 pupils**.

OR 120 ÷ 30 = 4, 4 × 10 = **40 pupils**.

[2 marks available — 1 mark for showing that 1/3 Year 9 pupils asked prefer fish and chips OR 1 mark for 120 ÷ 30 = 4; 1 mark for the correct final answer.]

Know your TWO-WAY TABLES to ace this question.

4 Time available = 6 × 60 = 360 minutes

total time for gaps = 15 × 5 = 75 minutes

360 − 75 = 285 minutes total time for 6 interviews

so maximum interview time = 285 ÷ 6 = **47.5 minutes**.

[3 marks available — 1 mark for showing '360 − 15 × 5' or '285'; 1 mark for attempting to find 285 ÷ 6; 1 mark for correct final answer.]

Check your final answer sounds sensible — you wouldn't interview each person for two days, or something daft like that.

5 a) 80p : 200p : 160p : 120p = 8 : 20 : 16 : 12 = **2 : 5 : 4 : 3**

[1 mark]

b) $\dfrac{42}{(2 + 5 + 4 + 3)} = \dfrac{42}{14} = 3$, so 1 part = £3

Sheena will get 2 parts = 2 × £3 = **£6.00**

Daryl will get 5 parts = 5 × £3 = **£15.00**

Helen will get 4 parts = 4 × £3 = **£12.00**

Ali will get 3 parts = 3 × £3 = **£9.00**

[4 marks available — 1 mark for finding that 1 part = £3; 1 mark for at least two correct amounts, 1 mark for at least 3 correct amounts, 1 mark for all amounts correct using the proportional division method shown above.]

It's PROPORTIONAL DIVISION — add up the parts, find the amount for one part, then find each amount.

6 a) $48 = 2 \times 2 \times 2 \times 2 \times 3$
   $48 = 2^4 \times 3$
   [2 marks available — 1 mark for finding 2 prime factors of 48;
   1 mark for correct final answer.]
   PRIME FACTORISATION — it's not funny.

 b) Multiples of 7 = 7, 14, 21, 28, 35, 42, 49, 56, 63, 70, 77, 84.
   Multiples of 12 = 12, 24, 36, 48, 60, 72, 84.
   LCM = **84**
   [2 marks available — 1 mark for finding the multiples
   of 7 and 12; 1 mark for the correct final answer.]

 c) Factors of 35 = 1, 5, 7, 35
   Factors of 63 = 1, 3, 7, 9, 21, 63.
   HCF = **7**
   [2 marks available — 1 mark for finding the factors of
   35 and 63; 1 mark for the correct final answer.]

7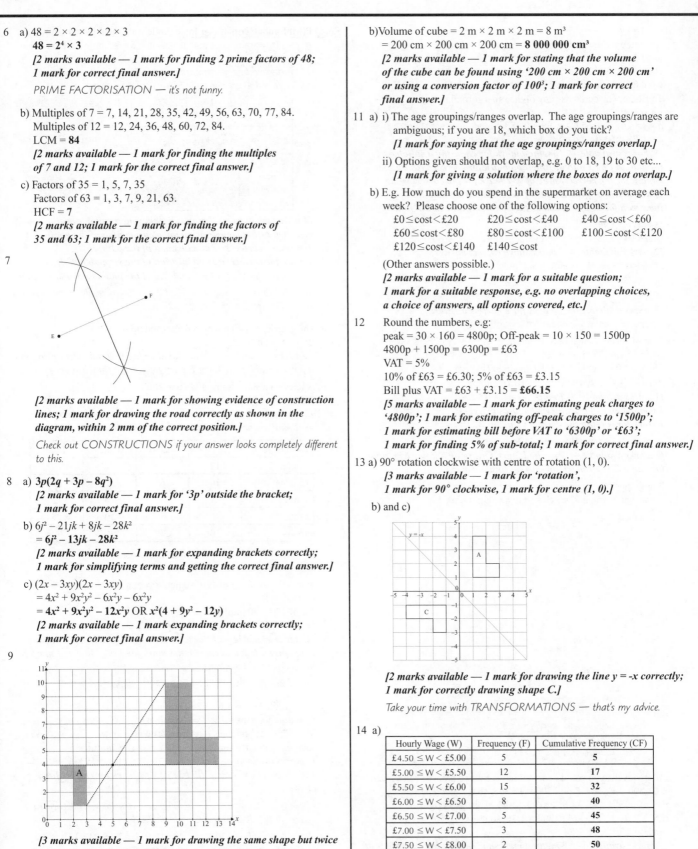

[2 marks available — 1 mark for showing evidence of construction
lines; 1 mark for drawing the road correctly as shown in the
diagram, within 2 mm of the correct position.]

Check out CONSTRUCTIONS if your answer looks completely different
to this.

8 a) **$3p(2q + 3p - 8q^2)$**
   [2 marks available — 1 mark for '3p' outside the bracket;
   1 mark for correct final answer.]

 b) $6j^2 - 21jk + 8jk - 28k^2$
   $= 6j^2 - 13jk - 28k^2$
   [2 marks available — 1 mark for expanding brackets correctly;
   1 mark for simplifying terms and getting the correct final answer.]

 c) $(2x - 3xy)(2x - 3xy)$
   $= 4x^2 + 9x^2y^2 - 6x^2y - 6x^2y$
   $= 4x^2 + 9x^2y^2 - 12x^2y$ OR $x^2(4 + 9y^2 - 12y)$
   [2 marks available — 1 mark expanding brackets correctly;
   1 mark for correct final answer.]

9

[3 marks available — 1 mark for drawing the same shape but twice
as big; 1 mark for drawing the shape correctly oriented;
1 mark for drawing the shape in the correct place using centre of
enlargement (5, 4).]

An ENLARGEMENT by -2 means the shape pops out the other side of
the enlargement centre, so that the shape is the other way round.

10 a) 6500 mm² = **65 cm²**
   [1 mark for correct final answer.]

 b) Volume of cube = 2 m × 2 m × 2 m = 8 m³
   = 200 cm × 200 cm × 200 cm = **8 000 000 cm³**
   [2 marks available — 1 mark for stating that the volume
   of the cube can be found using '200 cm × 200 cm × 200 cm'
   or using a conversion factor of 100³; 1 mark for correct
   final answer.]

11 a) i) The age groupings/ranges overlap. The age groupings/ranges are
      ambiguous; if you are 18, which box do you tick?
      [1 mark for saying that the age groupings/ranges overlap.]

   ii) Options given should not overlap, e.g. 0 to 18, 19 to 30 etc...
      [1 mark for giving a solution where the boxes do not overlap.]

 b) E.g. How much do you spend in the supermarket on average each
   week? Please choose one of the following options:
   £0 ≤ cost < £20   £20 ≤ cost < £40   £40 ≤ cost < £60
   £60 ≤ cost < £80   £80 ≤ cost < £100   £100 ≤ cost < £120
   £120 ≤ cost < £140   £140 ≤ cost

   (Other answers possible.)
   [2 marks available — 1 mark for a suitable question;
   1 mark for a suitable response, e.g. no overlapping choices,
   a choice of answers, all options covered, etc.]

12 Round the numbers, e.g:
   peak = 30 × 160 = 4800p; Off-peak = 10 × 150 = 1500p
   4800p + 1500p = 6300p = £63
   VAT = 5%
   10% of £63 = £6.30; 5% of £63 = £3.15
   Bill plus VAT = £63 + £3.15 = **£66.15**
   [5 marks available — 1 mark for estimating peak charges to
   '4800p'; 1 mark for estimating off-peak charges to '1500p';
   1 mark for estimating bill before VAT to '6300p' or '£63';
   1 mark for finding 5% of sub-total; 1 mark for correct final answer.]

13 a) 90° rotation clockwise with centre of rotation (1, 0).
   [3 marks available — 1 mark for 'rotation',
   1 mark for 90° clockwise, 1 mark for centre (1, 0).]

 b) and c)

[2 marks available — 1 mark for drawing the line y = -x correctly;
1 mark for correctly drawing shape C.]

Take your time with TRANSFORMATIONS — that's my advice.

14 a)

Hourly Wage (W)	Frequency (F)	Cumulative Frequency (CF)
£4.50 ≤ W < £5.00	5	**5**
£5.00 ≤ W < £5.50	12	**17**
£5.50 ≤ W < £6.00	15	**32**
£6.00 ≤ W < £6.50	8	**40**
£6.50 ≤ W < £7.00	5	**45**
£7.00 ≤ W < £7.50	3	**48**
£7.50 ≤ W < £8.00	2	**50**

[2 marks available — 1 mark for getting the first two entries correct;
1 mark for getting all 7 entries correct.]

b)

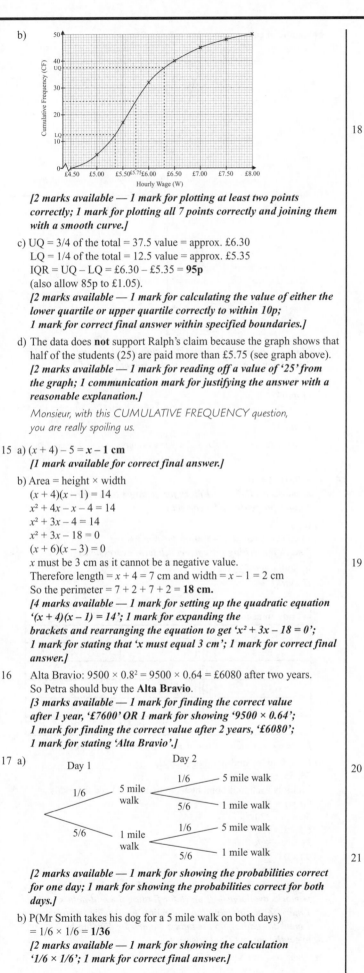

[2 marks available — 1 mark for plotting at least two points correctly; 1 mark for plotting all 7 points correctly and joining them with a smooth curve.]

c) UQ = 3/4 of the total = 37.5 value = approx. £6.30
LQ = 1/4 of the total = 12.5 value = approx. £5.35
IQR = UQ – LQ = £6.30 – £5.35 = **95p**
(also allow 85p to £1.05).
[2 marks available — 1 mark for calculating the value of either the lower quartile or upper quartile correctly to within 10p; 1 mark for correct final answer within specified boundaries.]

d) The data does **not** support Ralph's claim because the graph shows that half of the students (25) are paid more than £5.75 (see graph above).
[2 marks available — 1 mark for reading off a value of '25' from the graph; 1 communication mark for justifying the answer with a reasonable explanation.]

Monsieur, with this CUMULATIVE FREQUENCY question, you are really spoiling us.

15 a) $(x + 4) - 5 = $ **$x - 1$ cm**
[1 mark available for correct final answer.]

b) Area = height × width
$(x + 4)(x - 1) = 14$
$x^2 + 4x - x - 4 = 14$
$x^2 + 3x - 4 = 14$
$x^2 + 3x - 18 = 0$
$(x + 6)(x - 3) = 0$
x must be 3 cm as it cannot be a negative value.
Therefore length = $x + 4 = 7$ cm and width = $x - 1 = 2$ cm
So the perimeter = 7 + 2 + 7 + 2 = **18 cm.**
[4 marks available — 1 mark for setting up the quadratic equation '$(x + 4)(x - 1) = 14$'; 1 mark for expanding the brackets and rearranging the equation to get '$x^2 + 3x - 18 = 0$'; 1 mark for stating that 'x must equal 3 cm'; 1 mark for correct final answer.]

16 Alta Bravio: $9500 × 0.8^2 = 9500 × 0.64 = £6080$ after two years.
So Petra should buy the **Alta Bravio**.
[3 marks available — 1 mark for finding the correct value after 1 year, '£7600' OR 1 mark for showing '9500 × 0.64'; 1 mark for finding the correct value after 2 years, '£6080'; 1 mark for stating 'Alta Bravio'.]

17 a)
Day 1   Day 2

1/6 — 5 mile walk
1/6 — 5 mile walk
5/6 — 1 mile walk

5/6 — 1 mile walk
1/6 — 5 mile walk
5/6 — 1 mile walk

[2 marks available — 1 mark for showing the probabilities correct for one day; 1 mark for showing the probabilities correct for both days.]

b) P(Mr Smith takes his dog for a 5 mile walk on both days)
= 1/6 × 1/6 = **1/36**
[2 marks available — 1 mark for showing the calculation '1/6 × 1/6'; 1 mark for correct final answer.]

c) P(Mr Smith takes his dog for a 5 mile walk on at least 1 day)
= 1 – P(Mr Smith does not take his dog for a 5 mile walk on either day) = 1 – (5/6 × 5/6) = 1 – 25/36 = **11/36**
[2 marks available — 1 mark for showing the calculation '1 – (5/6 × 5/6)'; 1 mark for correct final answer.]

Got this one wrong? Then odds are you'll need to revise PROBABILITY.

18 a) $\sqrt{20} + \sqrt{45}$
$= \sqrt{4} \times \sqrt{5} + \sqrt{9} \times \sqrt{5}$
$= 2\sqrt{5} + 3\sqrt{5}$
$= \mathbf{5\sqrt{5}}$
[2 marks available — 1 mark for showing either 2√5 or 3√5; 1 mark for correct final answer.]

You're asked to simplify, not to use a calculator to give a rounded answer (which is a good job really since it's a non-calculator paper) — so remember to leave the square root sign in your answer.

b) $\dfrac{1}{\sqrt{3}}$
$= \dfrac{1 \times \sqrt{3}}{\sqrt{3} \times \sqrt{3}}$
$= \dfrac{\sqrt{3}}{3}$
[2 marks available — 1 mark for multiplying the numerator and denominator by √3; 1 mark for the correct final answer.]

c) $\dfrac{2 + \sqrt{3}}{2 - \sqrt{3}}$
$= \dfrac{(2 + \sqrt{3})(2 + \sqrt{3})}{(2 - \sqrt{3})(2 + \sqrt{3})}$
$= \dfrac{4 + 2\sqrt{3} + 2\sqrt{3} + 3}{4 - 3}$
$= \mathbf{7 + 4\sqrt{3}}$
[2 marks available — 1 mark for multiplying the numerator and denominator by 2 + √3; 1 mark for correct final answer.]

19 a) $8^{\frac{2}{3}} = \sqrt[3]{8^2} = \sqrt[3]{64} = \mathbf{4}$ OR $8^{\frac{2}{3}} = (\sqrt[3]{8})^2 = 2^2 = \mathbf{4}$
[2 marks available — 1 mark for squaring 8 or cube rooting 8; 1 mark for correct final answer.]

b) $4^{-3} = \dfrac{1}{4^3} = \mathbf{\dfrac{1}{64}}$
[2 marks available — 1 mark for writing $\dfrac{1}{4^3}$; 1 mark for correct final answer.]

c) $\left(\dfrac{27}{8}\right)^{-\frac{4}{3}} = \left(\dfrac{8}{27}\right)^{\frac{4}{3}} = \left(\dfrac{8^{\frac{4}{3}}}{27^{\frac{4}{3}}}\right) = \dfrac{(\sqrt[3]{8})^4}{(\sqrt[3]{27})^4} = \dfrac{2^4}{3^4} = \mathbf{\dfrac{16}{81}}$
[3 marks available — 1 mark for writing $\left(\dfrac{8}{27}\right)^{\frac{4}{3}}$; 1 mark for finding either '16' for the numerator or '81' for the denominator; 1 mark for final correct answer.]

20 a) Coordinates are: (30, 0.5) and (150, 0.5)
[2 marks available — 1 mark for finding each point.]

b) i) Curve C; $y = \sin(x + 180)$ OR $y = \cos(x + 90)$ OR $y = -\sin x$
[1 mark]
ii) Curve D; $y = 5 \sin x$
[1 mark]

21 a) **a – b**
[1 mark for correct final answer.]

b) **3a – 3b**
[1 mark for correct final answer.]

c) $\overrightarrow{MQ} = 3a - 3b = 3(a - b)$
$= 3 \times \overrightarrow{RS}$
Therefore RS is parallel to MQ.
[2 marks available — 1 mark for '3(a – b)'; 1 mark for correct final answer.]

Can't find the direction on this one — have a look at VECTORS.

22 $x + (x + 1) + (x + 2) + (x + 3) + (x + 4)$
$= 5x + 10$
$= 5(x + 2)$

OR $(x - 2) + (x - 1) + x + (x + 1) + (x + 2)$
$= 5x$

As the expression is divisible by 5 then the sum of any 5 consecutive numbers must also be divisible by 5.

[3 marks available — 1 mark for writing an expression that shows 5 consecutive numbers algebraically; 1 mark for simplifying this expression; 1 mark for factorising this expression and/or showing that the sum of 5 consecutive numbers must be divisible by 5.]

This is a SEQUENCES question in disguise... sneaky... The clues are 'five consecutive integers' (five terms in a row) and 'multiple of 5' (5x).

Set 3 Paper 2 — Calculator

1 a) 0.1875 *[1 mark]*

   b) $0.\dot{6}$ *[1 mark]*

   c) Let $x = 0.1\dot{8}$
   $100x = 18.1\dot{8}$
   $99x = 18.1\dot{8} - 0.1\dot{8} = 18$
   $x = \dfrac{18}{99} = \dfrac{2}{11}$

   [2 marks available – 1 mark for showing a correct method; 1 mark for the final correct answer.]

2 a) $4p + 8 = 36$
   $4p = 28$
   $p = 7$
   [1 mark]

   b) $6q + 13 = 12q - 2$
   $13 + 2 = 12q - 6q$
   $15 = 6q$
   $q = 2.5$
   [2 marks available — 1 mark for '15 = 6q'; 1 mark for correct final answer.]

   c) $3(2r + 14) = 2(25 + 4r)$
   $6r + 42 = 50 + 8r$
   $6r - 8r = 50 - 42$
   $-2r = 8$
   $r = -4$
   [3 marks available — 1 mark for correctly expanding both brackets; 1 mark for getting −2r = 8 and 1 mark for correct final answer.]

   A nice question on SOLVING EQUATIONS for a bit of good, honest fun. You're welcome.

3 a)

x	-1	0	1	2	3
y	4	1	0	1	4

   [2 marks available — 1 mark for getting at least 2 values of x correct; 1 mark for getting all 4 values of x correct.]

   b)

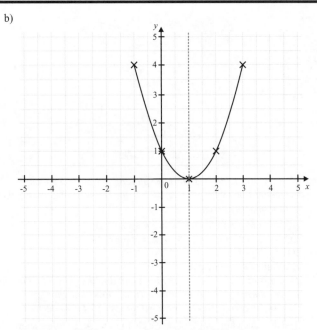

   [2 marks available — 1 mark for plotting all 5 values correctly; 1 mark for joining the points with a smooth curve.]

   c) i) A dotted line should be drawn as shown above.
   [1 mark]

   ii) $x = 1$
   [1 mark]

   Make sure you don't plot x and y the wrong way round when drawing QUADRATIC GRAPHS — it's a simple (and silly) mistake to make.

4 a) **4, 7, 10, 13, 16**
   [2 marks available — 1 mark for getting 3 terms correct; 1 mark for getting all 5 correct.]

   b) i) **5n + 3**
   [2 marks available — 1 mark for writing 5n; 1 mark for writing the correct nth term expression.]

   ii) 50th term = **253**
   [1 mark]

5 a)

Time in seconds	Tally	Frequency
48 – 52	I	1
53 – 57	IIII	4
58 – 62	III	3
63 – 67	IIII III	8
68 – 72	IIII	4
Total		20

   [2 marks available — lose 1 mark for each incorrect row.]

   b) First, find the mid-point of each class:
   50, 55, 60, 65, 70.
   Multiply each mid-point by its frequency and add the totals:
   $50 \times 1 = 50$
   $55 \times 4 = 220$
   $60 \times 3 = 180$
   $65 \times 8 = 520$
   $70 \times 4 = \underline{280}$
   $= 1250$
   Divide by the total, so $1250 \div 20 =$ **62.5 seconds.**
   [4 marks available — 1 mark for finding the mid-points; 1 mark for multiplying the mid-points by their frequencies; 1 mark for adding up the totals; 1 mark for dividing by 20 to get the correct final answer.]

   'Show your working' means you should write down every calculation you do. For example, in this question you can still get most of the marks as long as the examiner can see you know to use mid-points to work out the mean.

6 a) 15% of £27.50 = 0.15 × 27.5 = £4.125
£27.50 – 15% = 27.5 – 4.125 = 23.375 = **£23.38**
OR find 85% of £27.50 = 0.85 × 27.5 = **£23.38**
[2 marks available — 2 marks for the correct answer.
Otherwise 1 mark for '0.15 × 27.5' or '0.85 × 27.5'.]

b) 25% of £50 = 0.25 × 50 = 12.5, so 50 – 12.5 = £37.50
OR 75% of £50 = 0.75 × 50 = £37.50.

15% of £37.50 = 0.15 × 37.5 = 5.625,
so 37.5 – 5.625 = 31.875 = **£31.88**
OR 85% of £37.50 = 0.85 × 37.5 = **£31.88**
[3 marks available — 1 mark for showing either '0.25 × 50' and
'0.15 × 37.5' OR '0.75 × 50' and '0.85 × 37.5' in your calculations,
1 mark for correctly calculating the sale price, '£37.50', 1 mark for
the correct final answer.]

Take your time and do this PERCENTAGES question in steps
— don't try to find the answer by just doing one calculation.

7 Let h = diagonal distance between the two trees.
Difference in height between the two trees = 3.8 – 1.3 = 2.5 m
Using Pythagoras' theorem, $h^2 = 2.5^2 + 7.5^2 = 62.5$
$h = \sqrt{62.5}$
$h = 7.9$ m (to the nearest 10 cm)
Need to add 70 cm to either end for securing the rope, so the total
length = 7.9 + 0.7 + 0.7 = **9.3 m** (to the nearest 10 cm).
[4 marks available — 1 mark for '2.5 m'; 1 mark for correctly using
Pythagoras' theorem; 1 mark for correctly working out the value of
h; 1 mark for final answer correct to 1 decimal place.]

8 a) ½ × 5x × 7x = 63 cm²
$$\frac{35}{2}x^2 = \textbf{63 cm}^2$$
[2 marks available — 1 mark for stating '½ × 5x × 7x = 63 cm²';
1 mark for the correct final answer.]

b) $35x^2 = 126$ cm²
$x^2 = 126/35$
$x^2 = 3.6$
$x = 1.897...$
$5x = $ **9.5 cm (1 d.p.)**
[3 marks available — 1 mark for showing either 'x² = 126/35'
or 'x² = 3.6'; 1 mark for finding the value of 'x';
1 mark for the correct final answer.]

Don't forget the last step of your answer here — to find the height of the
triangle (5x). Always check your answer matches what the questions asks.

9 a) P(*Brand X* sale) = 4/60 = 1/15.
P(no *Brand X* sale) = 1 – P(*Brand X* sale) = 1 – 1/15 = **14/15**.

[2 marks available — 1 mark for stating P(Brand X sale) = 1/15 or
P(no Brand X sale) = 56/60 or equivalent; 1 mark for correct final
answer.]

b) In one day Leo earns £7.25 × 8 hours = £58.
In 22 working days Leo will earn £58 × 22 = £1276.
Number of expected bonus payments = 1/15 × 22
= 1.466... × £1.50 = £2.20.
Expected earnings, including bonus = £1276 + £2.20 = **£1278.20**
[3 marks available — 1 mark for giving the amount earned in
22 days as '£1276'; 1 mark for calculating the total bonus payment
of '£2.20'; 1 mark for the correct final answer.]

10 a)

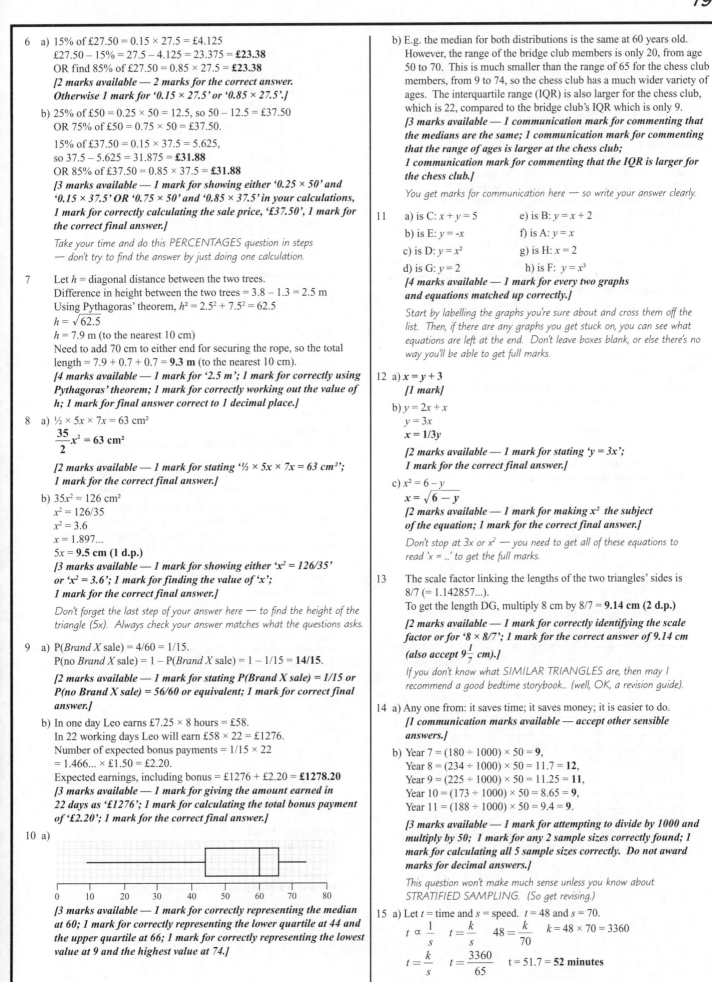

[3 marks available — 1 mark for correctly representing the median
at 60; 1 mark for correctly representing the lower quartile at 44 and
the upper quartile at 66; 1 mark for correctly representing the lowest
value at 9 and the highest value at 74.]

b) E.g. the median for both distributions is the same at 60 years old.
However, the range of the bridge club members is only 20, from age
50 to 70. This is much smaller than the range of 65 for the chess club
members, from 9 to 74, so the chess club has a much wider variety of
ages. The interquartile range (IQR) is also larger for the chess club,
which is 22, compared to the bridge club's IQR which is only 9.
[3 marks available — 1 communication mark for commenting that the
medians are the same; 1 communication mark for commenting
that the range of ages is larger at the chess club;
1 communication mark for commenting that the IQR is larger for
the chess club.]

You get marks for communication here — so write your answer clearly.

11 a) is C: $x + y = 5$ e) is B: $y = x + 2$
b) is E: $y = -x$ f) is A: $y = x$
c) is D: $y = x^2$ g) is H: $x = 2$
d) is G: $y = 2$ h) is F: $y = x^3$
[4 marks available — 1 mark for every two graphs
and equations matched up correctly.]

Start by labelling the graphs you're sure about and cross them off the
list. Then, if there are any graphs you get stuck on, you can see what
equations are left at the end. Don't leave boxes blank, or else there's no
way you'll be able to get full marks.

12 a) $x = y + 3$
[1 mark]

b) $y = 2x + x$
$y = 3x$
$x = 1/3y$
[2 marks available — 1 mark for stating 'y = 3x';
1 mark for the correct final answer.]

c) $x^2 = 6 - y$
$x = \sqrt{6 - y}$
[2 marks available — 1 mark for making x² the subject
of the equation; 1 mark for the correct final answer.]

Don't stop at 3x or x² — you need to get all of these equations to
read 'x = ...' to get the full marks.

13 The scale factor linking the lengths of the two triangles' sides is
8/7 (= 1.142857...).
To get the length DG, multiply 8 cm by 8/7 = **9.14 cm (2 d.p.)**
[2 marks available — 1 mark for correctly identifying the scale
factor or for '8 × 8/7'; 1 mark for the correct answer of 9.14 cm
(also accept $9\frac{1}{7}$ cm).]

If you don't know what SIMILAR TRIANGLES are, then may I
recommend a good bedtime storybook... (well, OK, a revision guide).

14 a) Any one from: it saves time; it saves money; it is easier to do.
[1 communication marks available — accept other sensible
answers.]

b) Year 7 = (180 ÷ 1000) × 50 = **9**,
Year 8 = (234 ÷ 1000) × 50 = 11.7 = **12**,
Year 9 = (225 ÷ 1000) × 50 = 11.25 = **11**,
Year 10 = (173 ÷ 1000) × 50 = 8.65 = **9**,
Year 11 = (188 ÷ 1000) × 50 = 9.4 = **9**.

[3 marks available — 1 mark for attempting to divide by 1000 and
multiply by 50; 1 mark for any 2 sample sizes correctly found; 1
mark for calculating all 5 sample sizes correctly. Do not award
marks for decimal answers.]

This question won't make much sense unless you know about
STRATIFIED SAMPLING. (So get revising.)

15 a) Let t = time and s = speed. $t = 48$ and $s = 70$.
$$t \propto \frac{1}{s} \quad t = \frac{k}{s} \quad 48 = \frac{k}{70} \quad k = 48 \times 70 = 3360$$
$$t = \frac{k}{s} \quad t = \frac{3360}{65} \quad t = 51.7 = \textbf{52 minutes}$$

[3 marks available — 1 mark for stating 't = k/s'; 1 mark for
working out 'k = 3360'; 1 mark for the correct final answer.]

b) i) Braking distance = $k \times$ (speed)^2.
[1 mark]

ii) If the speed doubles then the braking distance will increase by a factor of four.
[1 mark]

DIRECT AND INVERSE PROPORTION are really quite similar — but that's no excuse not to learn both of them.

16 a) $60/360 = 1/6$, so the sector is 1/6 the size of the full circle.
So the arc length ST is 1/6 of the circumference of the circle.
Arc length ST = $1/6 \pi d = 1/6 \pi \times 10 = $ **5.2 cm** (1 d.p.).
[2 marks available; 1 mark for stating that the arc length ST is 1/6 of the circumference of the circle; 1 mark for final correct answer.]

b) The area of the sector RST will also be 1/6 of the circle area. Area of Sector RST = $1/6 \pi r^2 = 1/6 \pi \times 5^2 = $ **13.09 cm²** (2 d.p.).

[2 marks available — 1 mark for correctly giving the formula for the area of the sector; 1 mark for correctly substituting in the value of the radius and hence showing the area to be 13.09 cm².]

Arcs and sectors — not half as exciting as they sound. It's all covered under AREAS. And whilst you're there, it's worth brushing up your knowledge about CIRCLE GEOMETRY too.

17 a) $(x + 11)(x - 2) = 0$
$x = $ **-11** or $x = $ **2**

[2 marks available — 1 mark for correctly factorising; 1 mark for both correct solutions.]

b) $(3x - 2)(2x + 2) = 0$
Either $3x - 2 = 0$, so $x = $ **2/3**
Or $2x + 2 = 0$, so $x = $ **-1**

[2 marks available — 1 mark for correctly factorising; 1 mark for both correct solutions.]

c) $3x^2 - x - 7 = 0$
Substituting the values into the quadratic formula,

$$x = \frac{1 \pm \sqrt{(-1)^2 - (4 \times 3(-7))}}{2 \times 3}$$

$$x = \frac{1 \pm \sqrt{85}}{6}$$

$$x = \frac{1 + \sqrt{85}}{6} = 1.70 \text{ (3 s.f.) or } x = \frac{1 - \sqrt{85}}{6} = \text{-1.37 (3 s.f.)}$$

[3 marks available — 1 mark for correctly substituting the values into the quadratic formula; 1 mark for each correct answer.]

The QUADRATIC FORMULA isn't something that'll just come to you on the day — you've really got to put the effort in now and learn how and when to use it.

18 a)

Time (m) minutes	Frequency	Frequency Density
$10 < m \leq 13$	**12**	4
$13 < m \leq 15$	13	6.5
$15 < m \leq 16$	**8**	8
$16 < m \leq 18$	**11**	5.5
$18 < m \leq 20$	10	5
$20 < m \leq 25$	6	1.2

[2 marks available — 1 mark for getting at least 1 value correct; 1 mark for getting all 3 values correct.]

b)

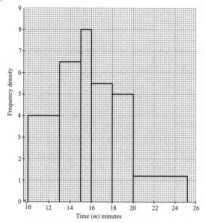

[2 marks available — 1 mark for drawing at least one bar to the correct height or showing on the table the correct values for the frequency density; 1 mark for correctly completing the histogram.]

c) $33/60 = $ **11/20**
[1 mark]

Oh look, it's another HISTOGRAM — they really are a favourite in exams, aren't they. Hmmm, could this possibly mean it'd be a good thing to revise them and make sure you know them inside out...?

19 a) 75.15 cm
[1 mark]

b) 75.25 cm
[1 mark; also accept 75.24999... to at least 3 d.p.]

c) $1.665 \div 0.7515 = 2.22$ m (2 d.p.).
[2 marks available — 1 mark for showing the values 1.665 and 0.7515; 1 mark for the correct final answer.]

d) Since the minimum value for the height of the door is $1.655 \div 0.7525 = 2.20$ m (2 d.p.), a value for the height of the door is 2.2 m (1 d.p.). This value satisfies both the maximum and minimum values calculated.

[2 marks available — 1 mark for finding minimum value of the door; 1 mark for suggesting 2.2 m to 1 d.p. for the height of the door.]

20 a) Volume of N = $1/3 \times \pi \times 9^2 \times 15 = 405 \pi$ cm³
Volume of P = $1/3 \times \pi \times 3^2 \times 5 = 15 \pi$ cm³
Therefore Volume of Q = $405 \pi - 15 \pi = 390 \pi$ cm³
[3 marks available — 1 mark for correctly showing the volume of N; 1 mark for correctly showing the volume of P; 1 mark for showing that the volume of N – P equals the volume of Q.]

If your answer doesn't equal 390π cm³ then you know you've gone wrong somewhere...

b) $390 \pi = 1/3 \times \pi \times r^2 \times 10$

$$\frac{390\pi}{\frac{1}{3} \times \pi \times 10} = r^2$$

$$\frac{1225.221135}{10.47197551} = r^2$$

$r = \sqrt{117}$
$r = $ **10.8 cm** (1 d.p.)
[3 marks available — 1 mark for writing the correct first line of working as shown above; 1 mark for rearranging the equation to make r² the subject; 1 mark for solving the equation and giving the correct value of r.]

MHP4U

GCSE Mathematics
Formula Sheet: Higher Tier

Area of trapezium $= \dfrac{1}{2}(a + b)h$

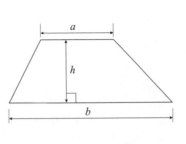

Volume of prism = area of cross-section × length

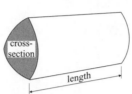

Volume of sphere $= \dfrac{4}{3}\pi r^3$

Surface area of sphere $= 4\pi r^2$

Volume of cone $= \dfrac{1}{3}\pi r^2 h$

Curved surface area of cone $= \pi r l$

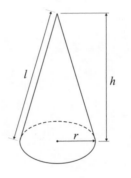

For any triangle ABC:

Sine rule: $\dfrac{a}{\sin A} = \dfrac{b}{\sin B} = \dfrac{c}{\sin C}$

Cosine rule: $a^2 = b^2 + c^2 - 2bc \cos A$

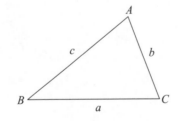

Area of triangle $= \dfrac{1}{2}ab \sin C$

The quadratic equation:

The solutions of $ax^2 + bx + c = 0$, where $a \neq 0$, are given by $x = \dfrac{-b \pm \sqrt{(b^2 - 4ac)}}{2a}$

General Certificate of Secondary Education

GCSE Mathematics

Practice Set 1

Paper 1 Non-calculator

Higher Tier

Time allowed: 1 hour 45 minutes

Centre name				
Centre number				
Candidate number				

Surname	
Other names	
Candidate signature	

In addition to this paper you should have:
- GCSE Mathematics Formula Sheet: Higher Tier.
- A ruler.
- A protractor.
- A pair of compasses.

Tracing paper may be used.

Instructions to candidates
- Write your name and other details in the spaces provided above.
- Answer **all** questions in the spaces provided.
- In calculations show clearly how you worked out your answers.

Information for candidates
- The marks available are given in brackets at the end of each question.
- You may get marks for method, even if your answer is incorrect.
- In questions labelled with an asterisk *, you will be assessed on the quality of your written communication — take particular care here with spelling, punctuation and the quality of explanations.

Advice to candidates
- Work steadily through the paper.
- Don't spend too long on one question.
- If you have time at the end, go back and check your answers.

1   Estimate the answer to the calculation below. Show your working.

$$\frac{30\,004 \times 0.102}{3.014 + 6.972}$$

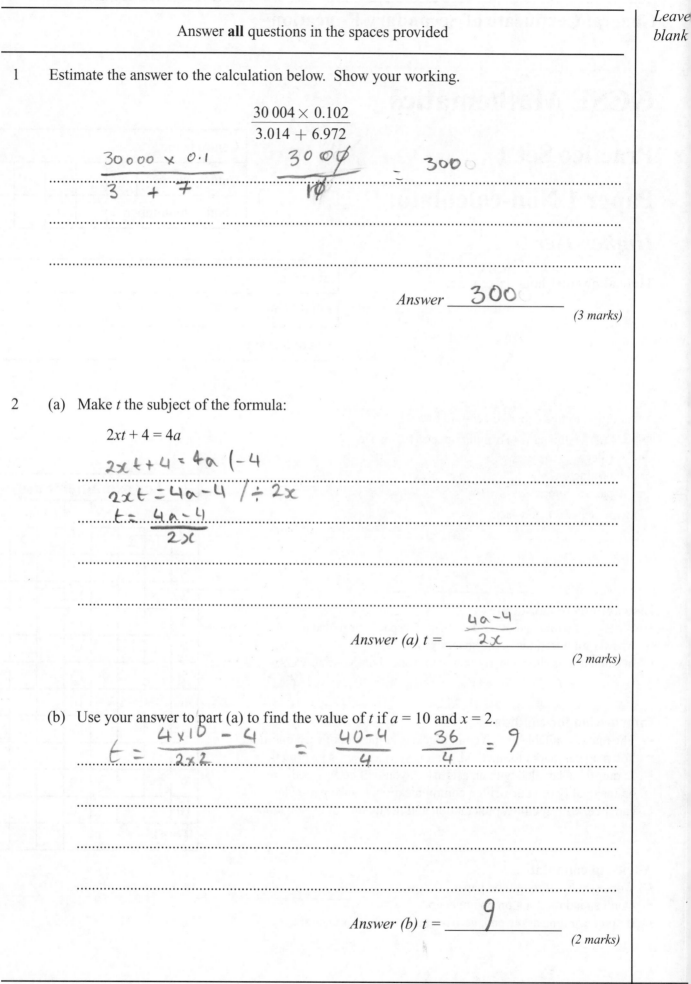

$$\frac{30000 \times 0.1}{3 + 7} \qquad \frac{3000}{10} = 3000$$

Answer   300

(3 marks)

2   (a)   Make *t* the subject of the formula:

$$2xt + 4 = 4a$$

$$2xt + 4 = 4a \; (-4$$

$$2xt = 4a - 4 \; / \div 2x$$

$$t = \frac{4a - 4}{2x}$$

Answer (a) t = $\dfrac{4a-4}{2x}$

(2 marks)

(b)   Use your answer to part (a) to find the value of *t* if *a* = 10 and *x* = 2.

$$t = \frac{4 \times 10 - 4}{2 \times 2} = \frac{40 - 4}{4} \quad \frac{36}{4} = 9$$

Answer (b) t = 9

(2 marks)

3   Dave is an estate agent. He is drawing an accurate scale plan of a house
     to be published in a property brochure. He is using a scale of 2 cm : 1 m.

(a)   The living room in the house is 5.5 m long by 4.2 m wide.
       What size should Dave draw the living room on his plan?

$5.5m = 11cm$ $4.2 = 84$

Answer (a) ____111____ cm by ____8.4____ cm

(2 marks)

(b)   Dave sketched the dining room when he took the house measurements, shown below.
       On the grid below, draw a scale plan of the dining room.
       Mark on all the correct lengths that Dave should use to draw his plan.

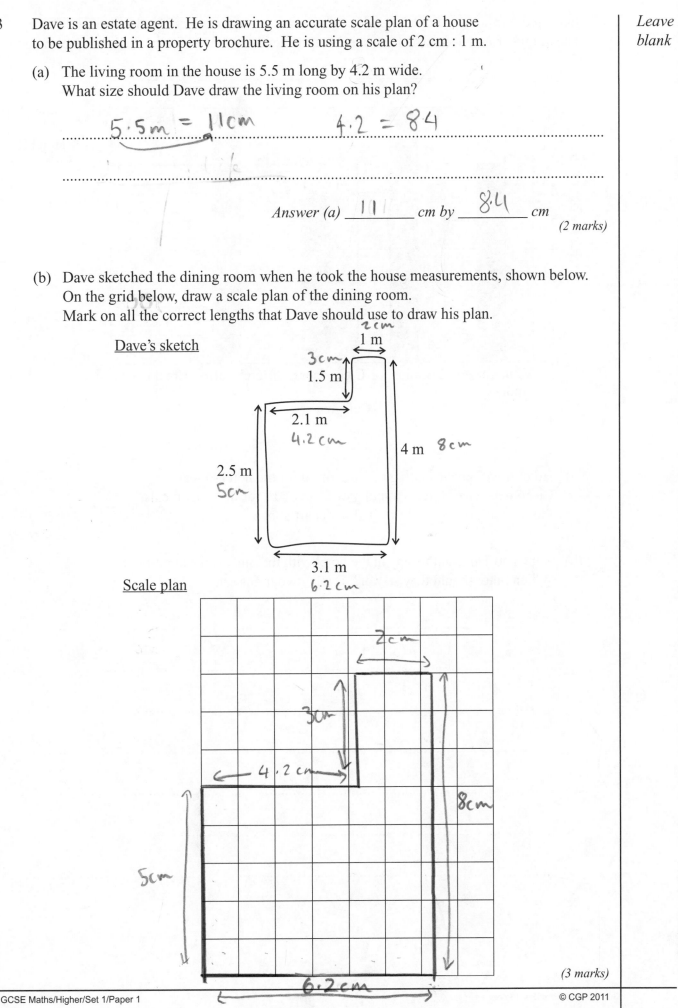

Dave's sketch

Scale plan

(3 marks)

GCSE Maths/Higher/Set 1/Paper 1

© CGP 2011

3

Leave blank

4   Jinny and Tim want to book a holiday. They want to stay in either Hotel A or Hotel B.
       The ratings from guests who have stayed at the hotels are shown in the dual bar chart below.

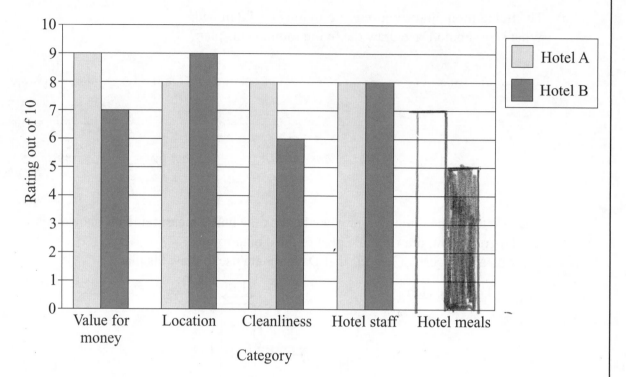

(a)   Hotel A was given a rating of 7 out of 10 for the hotel's meals.
      Customers gave Hotel B a rating of 5 out of 10 for the hotel meals.
      Plot this information on the dual bar chart above.

(1 mark)

(b)   Jinny and Tim want to stay in the hotel with the highest mean rating.
      Which hotel should they stay at? Explain your answer.

Hotel A = $9 + 8 + 8 + 8 + 7$ = $\frac{40}{5}$ = 8

Hotel B = $7 + 9 + 6 + 8 + 5$ = $\frac{35}{5}$ = 7

Hotel A.

(3 marks)

5   The pie charts show how classes 10H and 10Y did in a science module exam.

Class 10H   Class 10Y

(a) There are 36 students in Class 10H. How many pupils got a grade D or below?

$100° = D$ or below

$360 ÷ 10 = 36$ $100 ÷ 10 = 10$

Answer (a) ____10____ pupils

(2 marks)

A not higher proportion but greater number.

(b)* Donya says: "The pie charts show that <u>more people</u> got a grade A
    in Class 10H than 10Y". Is Donya right? Explain your answer.

Donya could be right but she cannot be
sure because we don't know how many
students are in 10y.

(4 marks)

6*   A primary school wants to find out whether their students enjoy school meals. The Headteacher gives out the questionnaire shown below.

> *Do you agree that school meals taste good?*
>
> ☐ **Yes**   ☐ **No**
>
> *Name* _____

(a)   Give two criticisms of this questionnaire.

The question is too closed. It should be an anonymous questionnaire

There is no option of 'sometimes' because the

lunches aren't always the same. There is no

'year', 'class' identification number.

(2 marks)

The Headteacher collects the pupils' questionnaires as they leave the school canteen.

(b)   Do you think this is a good method for the Headteacher to use? Explain your answer.

No because the children will have just eaten

so they may not remember other

(2 marks)

meals, (worse or better) also the

Headteacher may inhimidate the teacher

and make the questionaire

biased.

7   Work out the value of each expression below. Give your answers in standard form.

(a) $3 \times 10^{200} \times 2 \times 10^{100}$

6×10^{300}

Answer (a) 6×10^{300}

(2 marks)

(b) $3 \times 10^5 + 2 \times 10^4$

$3 \times 10^5 + 2 \times 10^4$ $10^5 \times 3.2$

$3 \times 10^4 \times 10^1 + 2 \times 10^4$

$3 \times 10^1 = 30 \times 10^4 + 2 \times 10^4$ Answer (b) 3.2×10^5

$10^4 (30 + 2) = 10^4 \times 32$

(2 marks)

(c) $(3 \times 10^{20}) \div (2 \times 10^{30})$

$\dfrac{3 \times 10^{20}}{2 \times 10^{30}}$ $\dfrac{3 \times 1}{2 \times 10^{10}}$ $\dfrac{3}{2 \times 10^{10}} = 1.5 \times 10^{-10}$

Answer (c) 1.5×10^{-11}

(2 marks)

8   Tom has a bag of red, green and yellow sweets.
    He picks a sweet out of the bag without looking. $P(R)$ $P(G)$ $P(y)$

The probability of him picking a red sweet is $\frac{1}{7}$.

The probability of him picking a green sweet is twice
the probability of him picking a yellow sweet.

(a)   Work out the probability of Tom picking a yellow sweet out of the bag.

$P(r) + P(g) + P(y) = 1$ $P(G) = 2 \times P(y)$

$\frac{1}{7} + 2P(y) + P(y) = 1$ $/ -\frac{1}{7}$

$3P(y) = \frac{6}{7}$ $/ \div 3$

*(See book
division of
fractions)*

$P(y) = \dfrac{\frac{6}{7}}{3} = \frac{2}{7}$ Answer (a) $\dfrac{2}{7}$

(3 marks)

(b)* Tom knows there are no more than 10 sweets in the bag.
      How many sweets must be in the bag? Explain your answer. $\frac{1}{7}$

If the probability is expressed in sevens then the
number of sweets must be a multiple of 7.

(2 marks)

7 is the only multiple of 7 in this
range.

9   In the diagram below, lines AC and ED are parallel. Lines BE and BD are the same length.

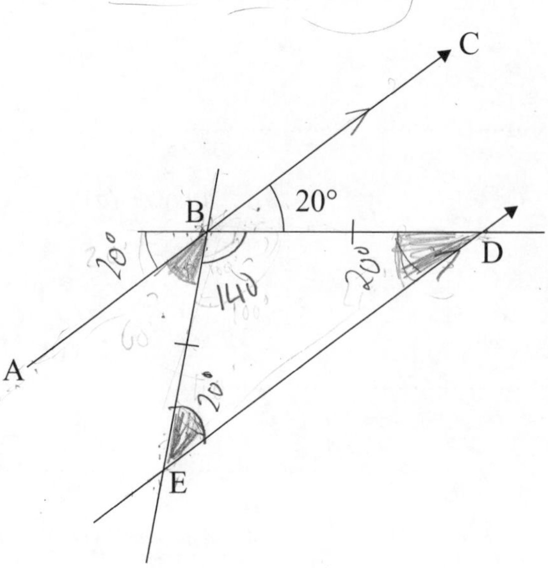

Not drawn to scale

Find angle ABE. Show all your working.

BED = BDE   (Isosceles triangle)

ABE = BED = 20° (Alternate angles)

EBD = 140° (180 - 20 - 20 , sum of angles in triangle = 180°)

Answer: Angle ABE = _____20_____ °

(3 marks)

10   (a)   Complete the table of values below for $y = \dfrac{60}{x+1}$.

x	0	1	2	3	4	5
$y = \dfrac{60}{x+1}$	60	30	20	15	12	10

(2 marks)

(b)   Use the table to draw the graph of $y = \dfrac{60}{x+1}$ on the graph paper below.

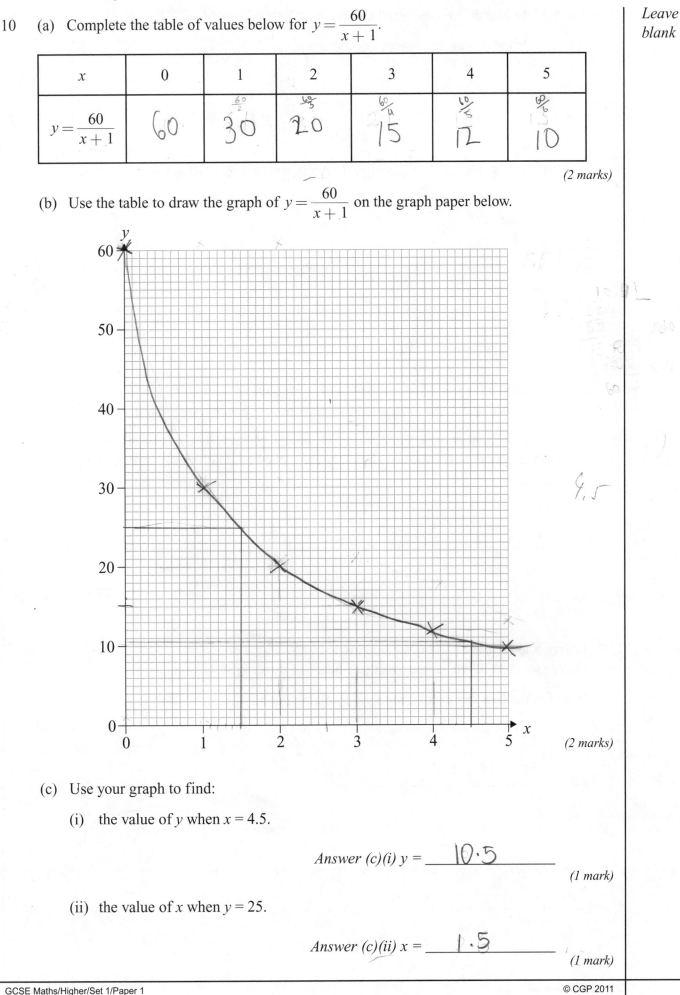

(2 marks)

(c)   Use your graph to find:

(i)   the value of y when $x = 4.5$.

Answer (c)(i) y = ___10.5___

(1 mark)

(ii)   the value of x when $y = 25$.

Answer (c)(ii) x = ___1.5___

(1 mark)

11 (a) (i) In March, Joy was taxed £44 on the £220 she earned.

What percentage rate of tax did she pay?

$$\frac{44}{220} = \frac{4}{20} = \frac{1}{5} : 20\%$$

Answer (a) (i) ___20___ %

(1 mark)

(ii) Joy is checking her pay slip. In April, she earned £310 before tax was deducted.

If the tax rate was the same as in part (i), how much should she have been paid after tax?

£310 − 20% of 310

1% = 3.1 × 20 = 62

310 - 62 =

62

248

Answer (a) (ii) £ ___248___

(2 marks)

don't pay

(b) Andy has a tax free allowance of £420 per month.
He is taxed at the same rate as Joy on any money he earns over £420.

Last month, Andy paid £130 in tax. He wants to check if he has been paid the correct amount. How much should he have been paid before tax was deducted?

130 in tax

$$\frac{130}{x} \times 100 = \frac{20}{100} \times 100 \quad \times 100$$

13000 = 20x

1300 = 2x

x = 650

taxable amount

650
+420 %
1070

Answer (b) £ ___1070.00___

(2 marks)

12* Darren and Kate are playing a game involving tossing 3 coins.
Darren says: "You win if you toss three heads or three tails. Otherwise, I win".

Who has the best chance of winning? Explain your answer.

Kate because: - Not clear if it has to be only heads or only tails.

(5 marks)

13   Hervine is designing a garden.
     She wants the shape of the lawn to be a sector
     of a circle, as shown in the diagram.

3 m

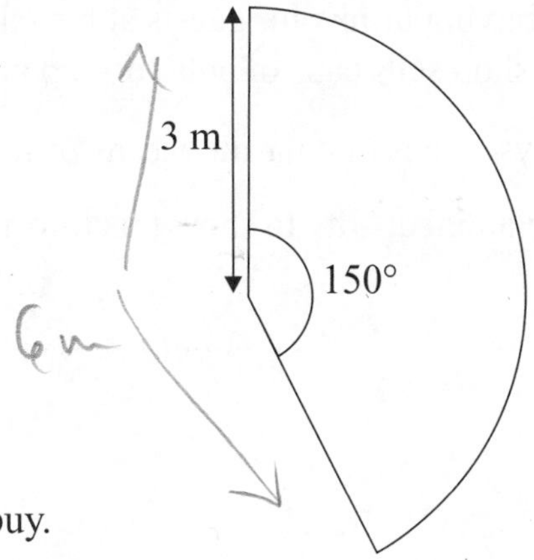

Not to scale

150°

6 m

Hervine decides to buy edging to go
around the perimeter of her lawn.

(a) Work out the length of edging she must buy.
     Give your answer in terms of π.

3m + 3m = 6m

$\pi 3^2 = \pi 3^2 - 210° + 6m$.

Answer (a) _____ m

(3 marks)

Hervine also needs to buy turf to cover the area of lawn.
Turf costs £28 per square metre.

(b) Work out the area of turf she needs to buy.
     Use the approximation $\pi \approx \dfrac{22}{7}$ to find out how much Hervine must pay for the turf.

Answer (b) £_____

(3 marks)

I can't remember inequalities. PLZ go over.

14 Danny is buying drinks for guests at his barbecue.
    The local shop sells cans of cola for 30p and cans of lemonade for 40p.

Danny buys x cans of cola and y cans of lemonade. He can spend a maximum of £6.

(a) Write an inequality to show this information.

....

Answer (a) _____

(1 mark)

(b) Danny has four guests who do not drink cola, so he must buy at least four cans of
    lemonade. Write down an inequality that describes how many cans of lemonade Danny
    must buy.

....

Answer (b) _____

(1 mark)

(c) Draw the inequalities from part (a) and part (b) on the grid below.
    Shade the region(s) that satisfy the inequalities.

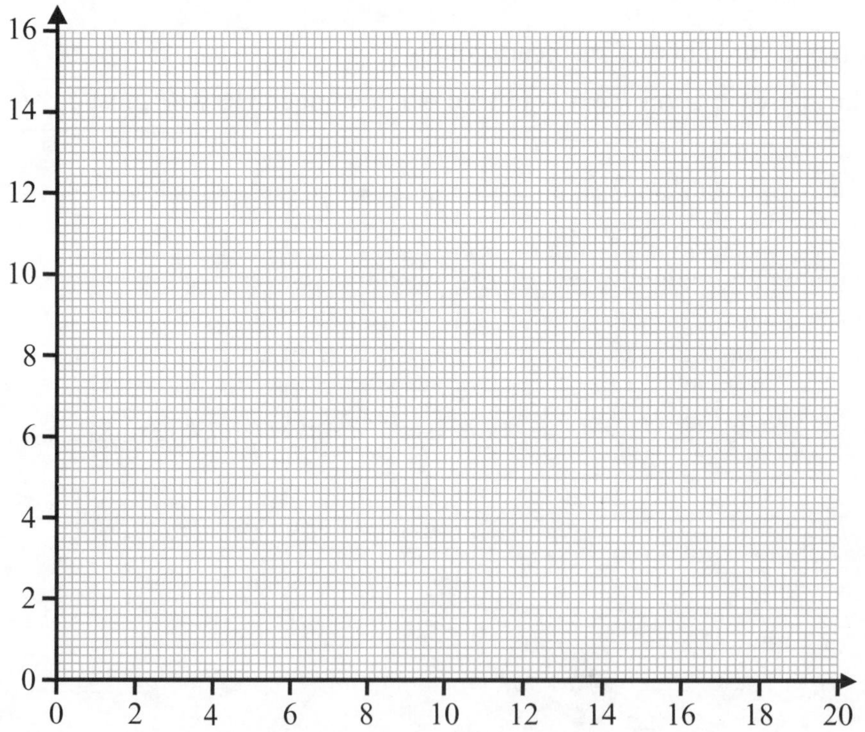

(5 marks)

(d) Danny buys 18 cans in total. State one combination of drinks he could have chosen.

....

(1 mark)

15 A regular polygon has an interior angle of 150°.

360° 150
       150
       300

    (a) Calculate how many sides this polygon has.

.... 300° = 2 sides. ....

....

....

                          Answer (a) _____ sides
                                                    (3 marks)

    (b) Explain why it is not possible to make a tessellating pattern using this polygon.

....

....

....
                                                    (2 marks)

16 (a) Write $2^3 \times (2^x)^4$ in its simplest form.

.... $2^3 \times (2^x)^4$ ....

.... $2^3 \times 2^{x+4}$ ....

2^{7+x}

                          Answer (a) ___ 2^{7+x} ___
                                                    (2 marks)

    (b) Solve the equation $2^{5+3x} = 2^2$.

.... $2^{5+3x} = 2^2$ ....

.... $2^{8x} = 2^2$ ....

$2^{8x} = 4$ / $^{-8}$

.... $2^x = 4^{-8}$ ....

                                        0.25
                          Answer (b) $x =$ ___ 2 ___
                                                    (2 marks)
$2^x = \frac{1}{4}$ / $\div 2$

$x = \frac{0.25}{2}$

13

17   A line has the equation $y = -2x + 3$.

(a) Write down the equation of the line perpendicular to $y = -2x + 3$, which goes through the point (0, 8).

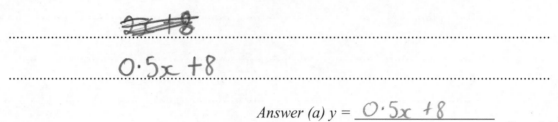

....

......................$0.5x + 8$..

Answer (a) $y = $ ___$0.5x + 8$___

(2 marks)

(b) A triangle is formed from the lines $y = -2x + 3$, the perpendicular line from part (a) and the *y*-axis.

Sketch the triangle on the grid below and find its area.

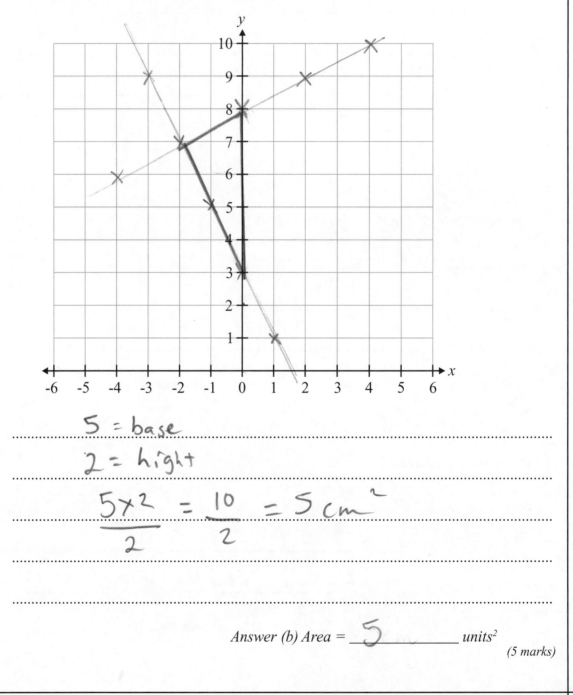

.................$5 = base$...

.................$2 = hight$...

.................$\dfrac{5 \times 2}{2} = \dfrac{10}{2} = 5\,cm^2$..

....

....

Answer (b) Area = ___5___ *units²*

(5 marks)

18 The diagram below shows the graph of $y = f(x)$.

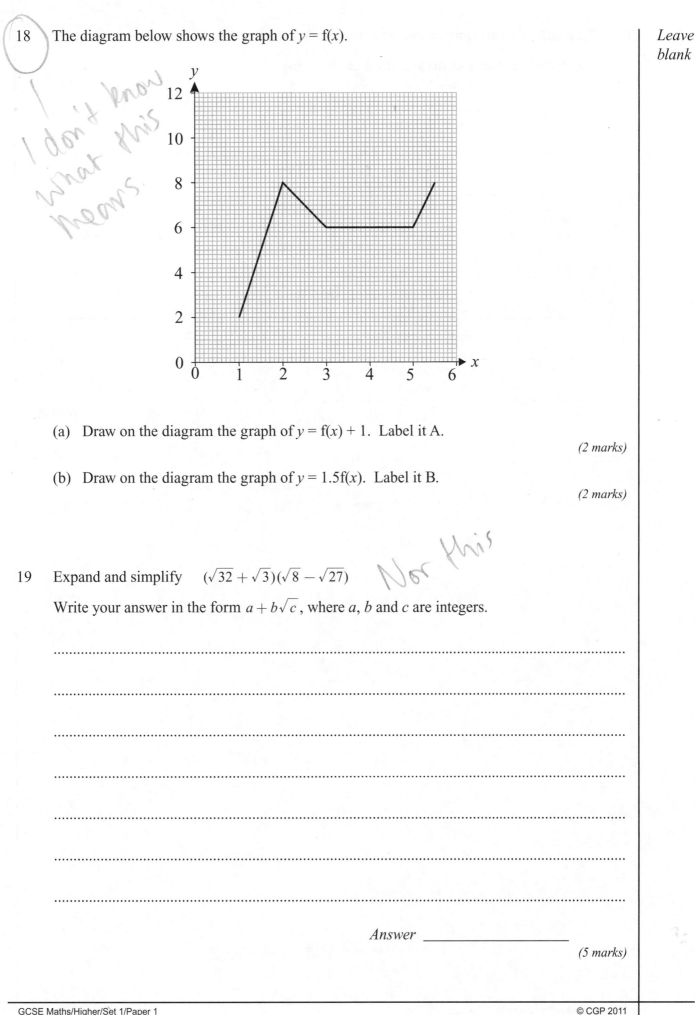

I don't know what this means.

(a) Draw on the diagram the graph of $y = f(x) + 1$. Label it A.

(2 marks)

(b) Draw on the diagram the graph of $y = 1.5f(x)$. Label it B.

(2 marks)

19 Expand and simplify $(\sqrt{32} + \sqrt{3})(\sqrt{8} - \sqrt{27})$

Nor this

Write your answer in the form $a + b\sqrt{c}$, where a, b and c are integers.

....

....

....

....

....

....

....

Answer _____

(5 marks)

20* The result of squaring an odd number and then subtracting 1 must always be divisible by 4.

Use <u>algebra</u> to show that the statement above is true.

not
Sure
how
to do
this.

$$\text{odd}^2 - 1 = \div 4$$

$$3^2 - 1 = 8$$

$$8 \div 4 = 2$$

(5 marks)

General Certificate of Secondary Education

GCSE Mathematics

Practice Set 1

Paper 2 Calculator

Higher Tier

Time allowed: 1 hour 45 minutes

Centre name				
Centre number				
Candidate number				

Surname	
Other names	
Candidate signature	

In addition to this paper you should have:
- GCSE Mathematics Formula Sheet: Higher Tier.
- A calculator.
- A ruler.
- A protractor.
- A pair of compasses.

Tracing paper may be used.

For examiner's use							
Q	Attempt Nº			Q	Attempt Nº		
	1	2	3		1	2	3
1				12			
2				13			
3				14			
4				15			
5				16			
6				17			
7				18			
8				19			
9				20			
10				21			
11							
Total							

Instructions to candidates
- Write your name and other details in the spaces provided above.
- Answer **all** questions in the spaces provided.
- In calculations show clearly how you worked out your answers.
- Take the value of π to be 3.142, or use the π button on your calculator.

Information for candidates
- The marks available are given in brackets at the end of each question.
- You may get marks for method, even if your answer is incorrect.
- In questions labelled with an asterisk *, you will be assessed on the quality of your written communication — take particular care here with spelling, punctuation and the quality of explanations.

Advice to candidates
- Work steadily through the paper.
- Don't spend too long on one question.
- If you have time at the end, go back and check your answers.

1   Harminder is thinking about changing his mobile phone contract.
     He sees the following adverts.

Contract A

Contract B

£20 a month

*Includes unlimited
texts and 100 minutes
(any extra calls are
10p a minute).*

£25 a month

*Includes 200 texts and
200 minutes (extra calls
are 15p a minute, extra
texts are 10p each).*

In an average month, Harminder makes 200 minutes of calls and sends 300 texts.
Work out which contract Harminder should choose. Show all your working.

....

....

....

....

Answer _____

(3 marks)

2   Dima is making a cylindrical draft excluder. She wants the length of the cylinder to be
     1.2 m. She has made circles with a diameter of 16 cm to make the ends of the cylinder.

     The filling she wants to use is sold in bags of 1 ft^3 or 2 ft^3. She knows that 1 ft$^3 \approx 0.03$ m^3.

     She wants to buy the smallest amount of filling possible.

     Which bag of filling should she buy? Show your working.

....

....

....

....

Answer _____

(3 marks)

3   Jenny and Danzeela want to estimate how often it rains in the UK. Jenny records how often it rains where she lives over one year. She found it rained on 220 out of 365 days.

Danzeela found records of UK rainfall on the internet for the last 10 years.
The website showed it had rained on 41% of the days.

   (a) Whose data gives the higher estimate of how often it rains, Jenny's or Danzeela's?
        Show all your working.

        ....

        ....

        Answer (a) _____

        (2 marks)

   (b)* Which estimate is more reliable? Explain your answer.

        ....

        ....

        ....

        (3 marks)

4   (a) Expand and simplify $(a-3)(a+4)$

        ....

        ....

        Answer (a) _____

        (2 marks)

   (b) Factorise fully $3xy - 9x^2y$

        ....

        ....

        Answer (b) _____

        (2 marks)

   (c) Factorise $x^2 + 6x + 8$

        ....

        ....

        Answer (c) _____

        (2 marks)

5   Look at the sequence below.

  5   8   11   14   17 ...

  (a) Find the next two terms of the sequence.

  ....

                                    Answer (a) _____
                                                        (1 mark)

  (b) Find the nth term of the sequence.

  ....

  ....

                                    Answer (b) _____
                                                        (2 marks)

  (c) Find the first term in the sequence that has a value larger than 100.

  ....

  ....

  ....

                                    Answer (c) _____
                                                        (3 marks)

6   A school asked 20 pupils in a Year 7 class how long they spend doing their homework each night, on average. The times, in minutes, are shown below.

40,   60,   85,   70,   80,   100,   35,   50,   60,   90,

90,   125,   45,   60,   65,   40,   120,   90,   85,   120.

(a) Draw a stem and leaf diagram to show this information in the space below.

(3 marks)

(b) Year 7 pupils have a test tomorrow. They all spend 30 minutes revising for the test on top of their usual homework. Work out the median time a Year 7 pupil would spend doing their homework and revision.

....

....

Answer (b) _____ *minutes*

(2 marks)

7   Vinod wants to buy a new MP3 player.
     He looks on three different websites and finds these offers:

     •   Website A: £80 + 20% VAT + £6 postage

     •   Website B: £100 including VAT. Free postage

     •   Website C: £120 including VAT. Free postage. 20% reduction if you buy now!

     Which website should Vinod buy an MP3 player from to get the cheapest deal?
     Show your working.

     ....

     ....

     ....

     ....

                                                          Answer _____
                                                                              (3 marks)

8   (a) Solve $4q + 9 = 1$

          ....

          ....

                                                     Answer (a) q = _____
                                                                              (2 marks)

     (b) Solve $y^3 - 18 = 36 - y^3$

          ....

          ....

          ....

                                                     Answer (b) y = _____
                                                                              (3 marks)

9   Every Saturday, Shannon either goes to the cinema, goes out for a pizza or visits her friend.

She goes out for a pizza 10% of the time, and goes to the cinema ¼ of the time.

(a)   On a randomly chosen Saturday, what is the probability Shannon visits her friend?
Give your answer as a decimal.

....

....

Answer (a) _____

(2 marks)

(b)   What is the probability that Shannon goes out for a pizza both this Saturday
and next Saturday?

....

....

Answer (b) _____

(2 marks)

(c)   Shannon keeps a diary of which activity she does each Saturday for a
long period of time. She finds she has gone to the cinema 15 times.

How many weeks has Shannon been keeping her diary for?

....

Answer (c) _____ *weeks*

(1 mark)

10 Christine is overweight and has been advised to take up jogging.
She wants to see if jogging extra miles makes a difference to her weight loss.
She weighs herself at the end of each week and has plotted a scatter graph to show
her weight loss in pounds (lbs) compared to how many miles she jogged that week.

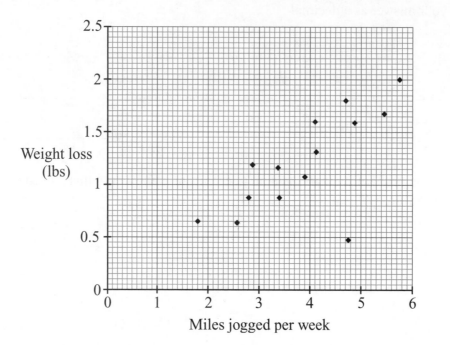

(a)* Describe the relationship between the distance Christine jogs and her weight loss.

....

....

....

(2 marks)

(b) Draw a line of best fit for the data.

(1 mark)

(c) Christine needs to lose 1.75 lbs to meet her target weight by the end of the week.
Estimate how many miles Christine should jog during the week.

Answer (c) _____

(1 mark)

11   Ellie needs to lay a concrete floor in a room in her new house. Below is a plan of
      the room. The curved area of the room is a semicircle with a diameter of 2 m.

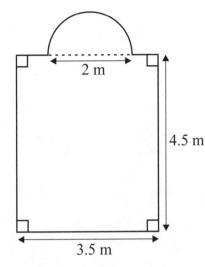

2 m

4.5 m

3.5 m

      Ellie needs to cover the entire floor with a 10 cm thick layer of concrete.
      Concrete costs £80 per cubic metre. Calculate how much it will cost
      Ellie to lay the concrete floor. Give your answer to the nearest pound.

      ....

      ....

      ....

      ....

                                                          *Answer £*_____
                                                                                (4 marks)

12   Solve these simultaneous equations:

$$3x + 4y = 43 \qquad\qquad 9x - 6y = 30$$

....

....

....

....

....

Answer x = _____

y = _____
(4 marks)

13   James measures the angle of elevation of the top of a tree to be 56°.

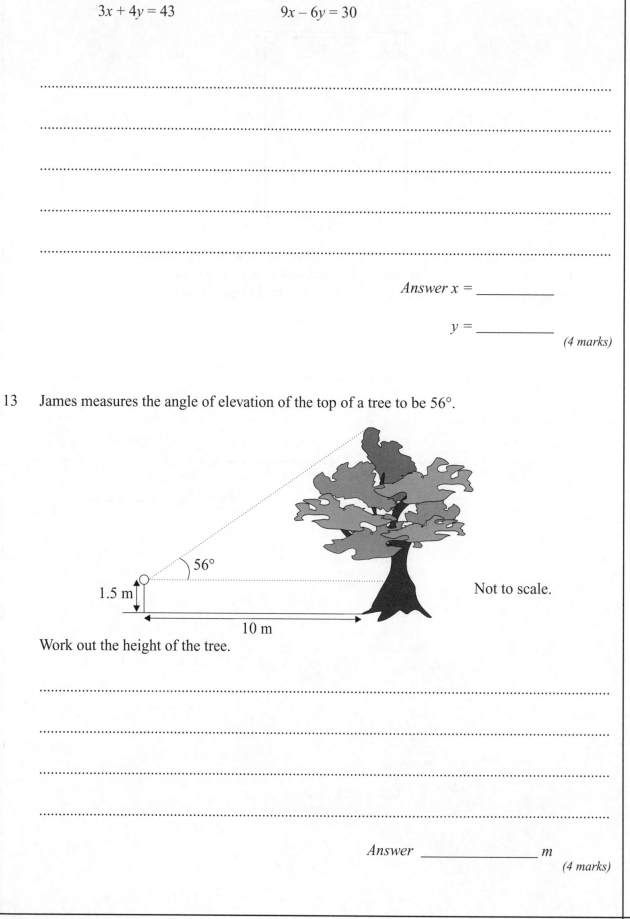

56°

1.5 m

10 m

Not to scale.

Work out the height of the tree.

....

....

....

....

Answer _____ *m*
(4 marks)

14 (a) A solution to $3x^3 - 2x - 64 = 0$ lies between 2.8 and 2.9.
Find the solution correct to 2 decimal places.

....

....

....

....

....

Answer (a) _____

(4 marks)

(b)* Allie says that a solution to $2x^3 + 3x = 64$ lies between 2 and 3.
Explain why she is incorrect. DO NOT FIND A SOLUTION.

....

....

....

Answer (b) _____

(2 marks)

15   Teenagers aged 14 to 17 were asked how much money on average they spend on new clothes every year. The results are shown in the histogram below.

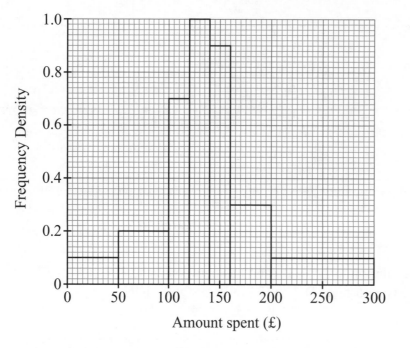

Amount spent (£)

(a)   Use the histogram to complete the frequency table below.

Amount spent on clothes per year (£)	Frequency
$0 \leq x < 50$	
$50 \leq x < 100$	
$100 \leq x < 120$	
$120 \leq x < 140$	
$140 \leq x < 160$	
$160 \leq x < 200$	
$200 \leq x < 300$	
$x \geq 300$	

(3 marks)

(b)   Use the frequency table to estimate the mean amount of money that the teenagers spend on new clothes per year.

....

....

....

*Answer (b) £*_____

(3 marks)

16   Faiza and Andrew want to buy a sofa which costs £1200.
      They can buy the sofa from either Shop A or Shop B.
      Both shops will allow Faiza and Andrew to delay paying for the sofa for 2 years.
      The shops' interest rates and delivery charges are shown below.

   •   Shop A charges 4% interest each year, with no repayments for 2 years and free delivery.

   •   Shop B only charges 3.7% interest each year, with no repayments for 2 years.
       However, Faiza and Andrew will have to pay £100 for delivery.

   Which shop would it be cheapest to buy the sofa from? Show all your working.

   ....

   ....

   ....

   ....

                                          Answer _____
                                                                        (4 marks)

17   Study the cuboid below.

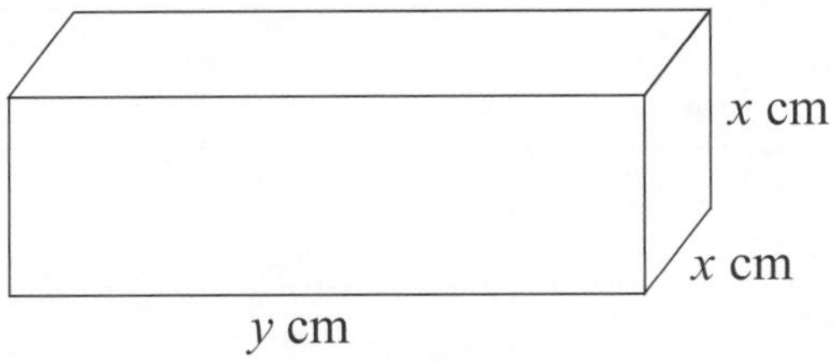

      x cm

      x cm

      y cm

                                    Not to scale.

   The length *x* is 40 cm to the nearest whole number.
   The volume of the cuboid is 80 000 cm³ correct to 4 significant figures.

   Work out the upper and lower bounds for length *y*.
   Give your answers correct to 3 significant figures.

   ....

   ....

   ....

   ....

   ....

                     Answer (upper bound) y = _____ *cm*

                            (lower bound) y = _____ *cm*
                                                                        (5 marks)

18   In the diagram below, C is the centre of the circle and PT is a tangent.

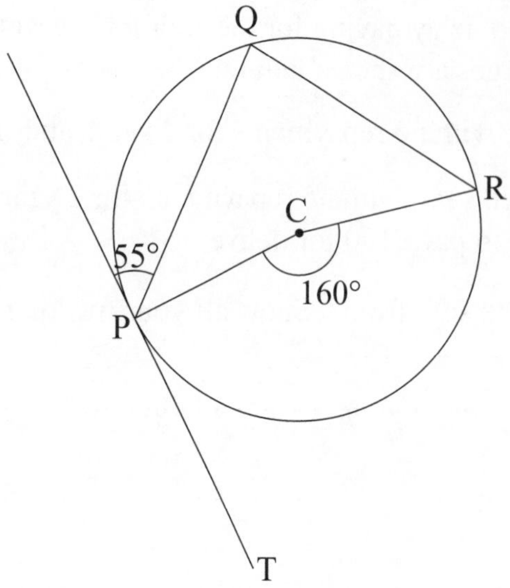

Not drawn accurately.

Find the size of angle CRQ. Show all of your working.

....

....

....

....

....

Answer: Angle CRQ = _____ °

(5 marks)

19   A triangle has a base length that is 4 cm longer than its height. It has a total area of 13 cm².
     Find the height of the triangle. Give your answer correct to three significant figures.

....

....

....

....

....

Answer _____ *cm*

(5 marks)

20   Zahra has bought her friend a variety of classic sweets for her birthday. Zahra wants
      to find a gift box that even the largest sweet, a 20 cm long stick of rock, will fit in.

      An online shop sells three different gift boxes, shown below.

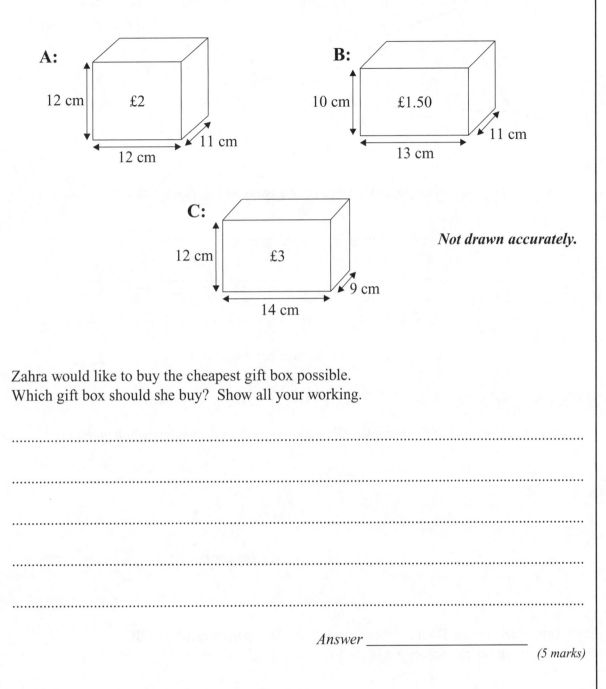

      Not drawn accurately.

      Zahra would like to buy the cheapest gift box possible.
      Which gift box should she buy? Show all your working.

      ....

      ....

      ....

      ....

      ....

                                        Answer _____
                                                          (5 marks)

21   Here is a triangle.

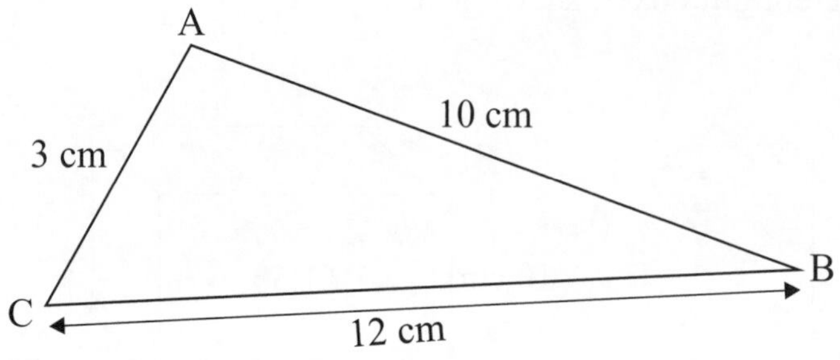

Not drawn accurately.

(a)   Find the size of angle BAC. Give your answer to 1 decimal place.

....

....

....

Answer (a) Angle BAC = _____ °

(3 marks)

(b)   Find the area of the triangle. Give your answer to the nearest whole number.

....

....

Answer (b) _____ *cm²*

(2 marks)

(c)   Mark a point D on the diagram so that AD is perpendicular to BC.
      Work out the distance AD.

....

....

Answer (c) AD = _____ *cm*

(2 marks)

General Certificate of Secondary Education

GCSE Mathematics

Practice Set 2
Paper 1 Non-calculator
Higher Tier

Time allowed: 1 hour 45 minutes

Centre name				
Centre number				
Candidate number				

Surname	
Other names	
Candidate signature	

In addition to this paper you should have:
- GCSE Mathematics Formula Sheet: Higher Tier.
- A ruler.
- A protractor.
- A pair of compasses.

Tracing paper may be used.

For examiner's use

Q	Attempt №			Q	Attempt №		
	1	2	3		1	2	3
1				12			
2				13			
3				14			
4				15			
5				16			
6				17			
7				18			
8				19			
9				20			
10				21			
11							
Total							

Instructions to candidates
- Write your name and other details in the spaces provided above.
- Answer **all** questions in the spaces provided.
- In calculations show clearly how you worked out your answers.

Information for candidates
- The marks available are given in brackets at the end of each question.
- You may get marks for method, even if your answer is incorrect.
- In questions labelled with an asterisk *, you will be assessed on the quality of your written communication — take particular care here with spelling, punctuation and the quality of explanations.

Advice to candidates
- Work steadily through the paper.
- Don't spend too long on one question.
- If you have time at the end, go back and check your answers.

Answer **all** questions in the spaces provided

1 (a) Calculate $\frac{11}{17} + \frac{25}{17}$. Give your answer as a mixed number.

....

....

Answer (a) _____

(2 marks)

(b) Simplify:

(i) $2\frac{4}{5} - \frac{7}{9}$

....

....

Answer (b) (i) _____

(2 marks)

(ii) $\frac{5}{6} \div 4\frac{11}{12}$

....

....

Answer (b) (ii) _____

(2 marks)

2 (a) Solve $10 - 3x > 5x - 6$.

....

....

Answer (a) _____

(1 mark)

(b) Show your solution to part (a) on the number line below.

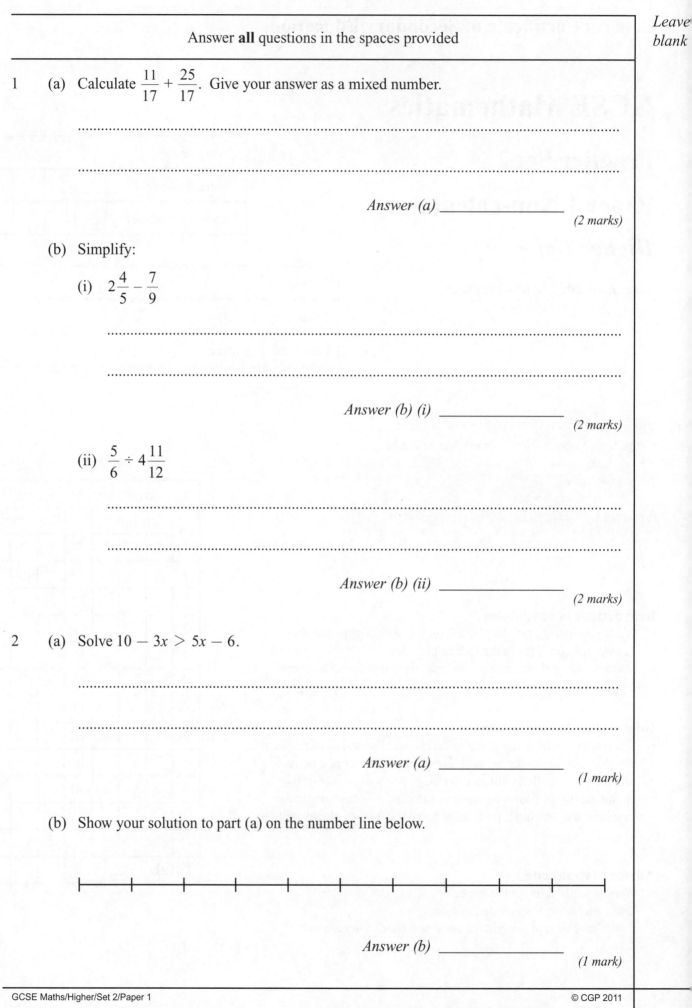

Answer (b) _____

(1 mark)

3   The locations of two towns, Acton and Blye, are shown below.

N
▲
|
●
Acton

N
▲
|
●
Blye

There are two other towns nearby:

Callow is due North of Blye, at a bearing of 070° from Acton.
Deane is due East of Acton, at a bearing of 068° from Blye.

(a)   On the diagram above, accurately mark the positions of Callow and Deane.

(2 marks)

(b)   What is the bearing of Deane from Callow?

Answer (b) _____ °

(1 mark)

(c)   The distance from Acton to Blye is 80 km. Find the distance from Blye to Callow.

....

....

Answer (c) _____ km

(2 marks)

4   Suzanne is going to university and is looking at accommodation prices for her first year,   *Leave*
     shown below. She can either pay in 3 equal termly instalments or in 8 monthly instalments,   *blank*
     from October to May.

Accommodation	Termly payments (3 equal payments)	Monthly payments (Oct-May)
Sea View	£1150	£432
The Maltings	£1200	£455
The Courtyards	£1170	£440

(a) If Suzanne pays termly, how much money will she save by spending
    her first year living in The Courtyards rather than The Maltings?

    ....

                                                          Answer (a) £ _____
                                                                                        (1 mark)

(b) Suzanne thinks she might like to live at Sea View.
    How much more will it cost her to pay monthly rather than termly to live there?

    ....

    ....

                                                          Answer (b) £ _____
                                                                                        (2 marks)

(c) There is another hall called Peregrine Heights. A twin room in Peregrine Heights costs
    £3300 per year. A single room in the same hall costs 30% more than a twin room.
    How much will it cost Suzanne to stay in a single room in Peregrine Heights?

    ....

    ....

                                                          Answer (c) £ _____
                                                                                        (2 marks)

5   Tanya wants to tile this floor:

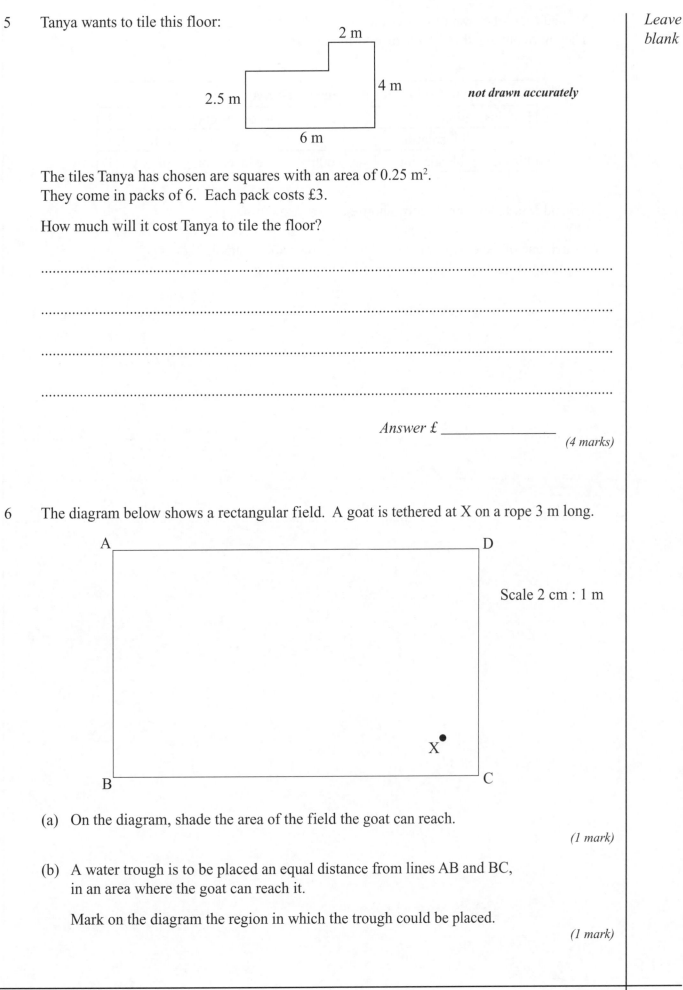

2 m

4 m

2.5 m

not drawn accurately

6 m

The tiles Tanya has chosen are squares with an area of 0.25 m².
They come in packs of 6. Each pack costs £3.

How much will it cost Tanya to tile the floor?

....

....

....

....

Answer £ _____

(4 marks)

6   The diagram below shows a rectangular field. A goat is tethered at X on a rope 3 m long.

A   D

Scale 2 cm : 1 m

X•

B   C

(a) On the diagram, shade the area of the field the goat can reach.

(1 mark)

(b) A water trough is to be placed an equal distance from lines AB and BC,
     in an area where the goat can reach it.

     Mark on the diagram the region in which the trough could be placed.

(1 mark)

7* Mr and Mrs Jones want to lower their energy bill from £1200.
They have put together a table of information, shown below.

Energy saving solution	Average cost	Average saving
Loft insulation	£75	Reduces energy bill by 1/8
Cavity wall insulation	£250	Reduces energy bill by 1/5
Plastic double glazing	£600	Reduces energy bill by 1/3

Mr and Mrs Jones want to move house in two years' time.

Which one of the energy saving solutions shown above should they buy?
Explain your answer. You must show all your working.

....

....

....

....

....

(5 marks)

8   Estimate the value of $\dfrac{9.56^2 - \sqrt{17.6}}{1.75^3}$.

....

....

....

Answer _____

(3 marks)

9   David is buying coffee and doughnuts for himself and some friends at work.
He buys 7 cups of coffee and 4 doughnuts for a total of £12.00.

When he gets out of the shop, he realises that he's forgotten two people's orders.
He goes back and buys another cup of coffee and 2 doughnuts for a total of £4.00.

Calculate the cost of 1 cup of coffee and the cost of 1 doughnut. Show your working.

....

....

....

....

....

....

Coffee: _____ *p, Doughnut: £* _____

(5 marks)

10 (a) Write 478 000 in standard form.

....

Answer (a) _____

(1 mark)

(b) Calculate $\dfrac{(4 \times 10^6) + (8 \times 10^7)}{(2 \times 10^3)}$

Give your answer in standard form.

....

....

....

Answer (b) _____

(3 marks)

11 Stuart is tiling his kitchen. The tiles he has bought are regular octagons.
He needs to buy some more tiles of a different shape to make a tessellating pattern.

By considering the sizes of the internal and external angles of regular octagons,
show that these tiles should be squares.

Sketch a section of the tessellation to illustrate your answer.

....

....

....

....

(4 marks)

12   Solve $\dfrac{4(x+3)}{2} - \dfrac{(x+2)}{3} = 3$

....

....

....

....

Answer x = _____

(4 marks)

13*   *Parkers Patios* are trialling a new telephone answering system.
      They have collected data on the time taken for the telephone to be answered with and
      without the system installed. The results are summarised in the box plots below.

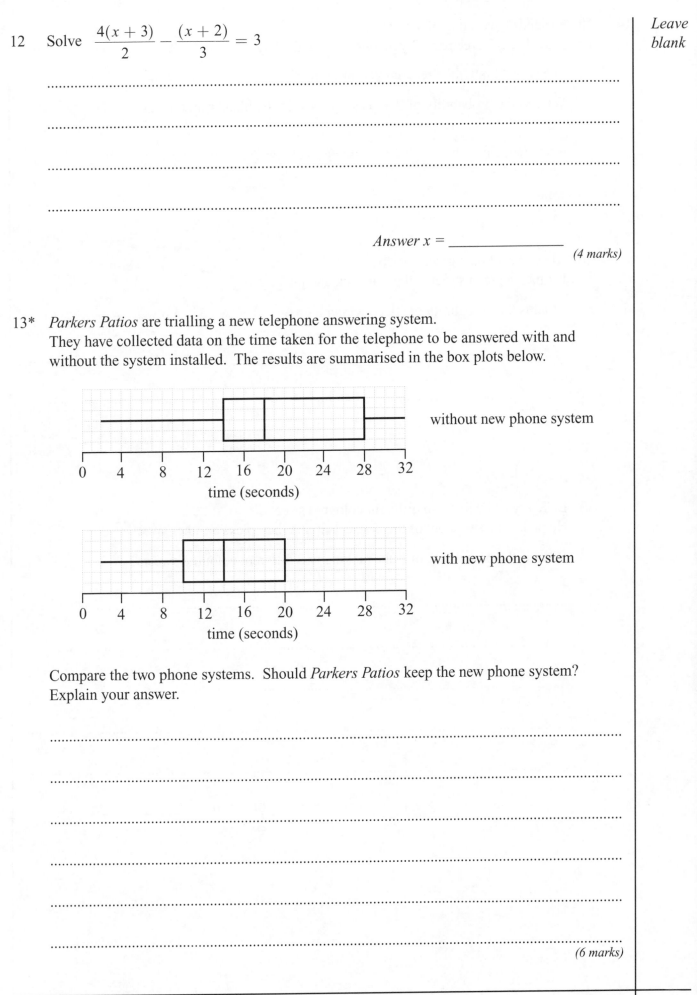

without new phone system

time (seconds)

with new phone system

time (seconds)

Compare the two phone systems. Should *Parkers Patios* keep the new phone system?
Explain your answer.

....

....

....

....

....

(6 marks)

14   Three friends have a bag of sweets.
      There are 4 red, 5 green and 3 yellow sweets in the bag.

Leave
blank

   (a) Sue says, "I only like red and yellow sweets."

        What is the probability of Sue picking a sweet she likes from the bag?

        ....

                                         Answer (a) _____
                                                                   (1 mark)

   (b) Alec says, "I hate green sweets.
        If I pick a green sweet, I'll put it back and pick again."

        What is the probability of Alec picking a green sweet twice in a row?

        ....

        ....

                                         Answer (b) _____
                                                                   (2 marks)

   (c) Luke says, "I'd like two different coloured sweets."
        He picks one sweet out of the bag, eats it and then picks another sweet.

        What is the probability of Luke picking two different coloured sweets?

        ....

        ....

        ....

        ....

                                         Answer (c) _____
                                                                   (4 marks)

GCSE Maths/Higher/Set 2/Paper 1

10

© CGP 2011

15 (a) Complete the table for $y = 3x^2 - 2$.

x	-3	-2	-1	0	1	2	3
$y = 3x^2 - 2$	25		1		1		25

(2 marks)

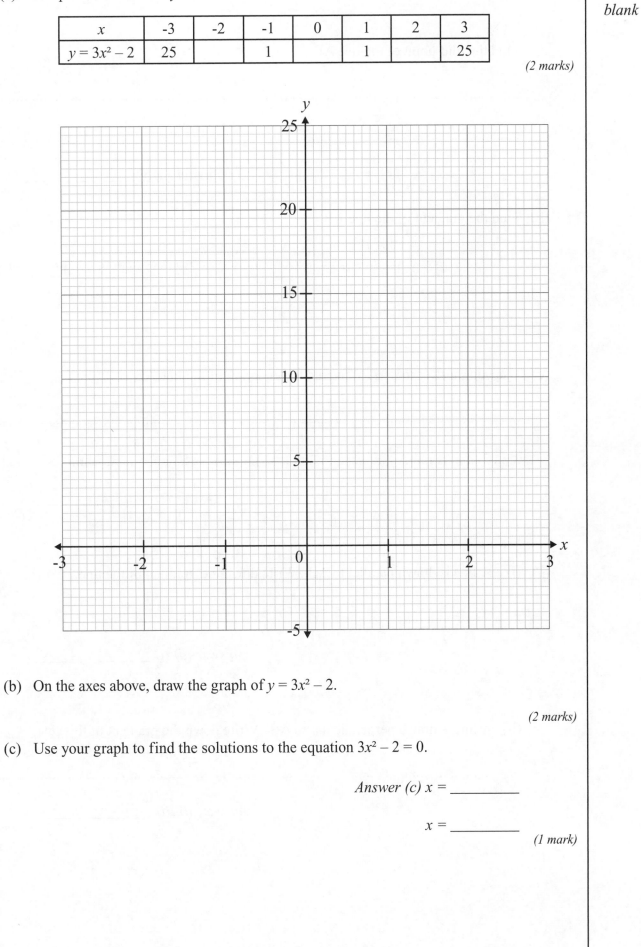

(b) On the axes above, draw the graph of $y = 3x^2 - 2$.

(2 marks)

(c) Use your graph to find the solutions to the equation $3x^2 - 2 = 0$.

Answer (c) x = _____

x = _____

(1 mark)

16   A line is drawn on a coordinate grid between point A (3, 9) and point B (8, 24).

Leave blank

(a) Find the mid-point of the line AB.

....

....

Answer (a) _____
(2 marks)

(b) Find the gradient of the line AB.

....

Answer (b) _____
(1 mark)

(c) Use your answer to part (b) to find the equation of the line AB.

....

....

Answer (c) _____
(2 marks)

(d) (i)   A line is drawn parallel to AB. Write down the gradient of the line.

....

Answer (d) (i) _____
(1 mark)

(ii) A line is drawn perpendicular to AB. Write down the gradient of this line.

....

Answer (d) (ii) _____
(1 mark)

17   In the diagram below, line AB is parallel to line ED.

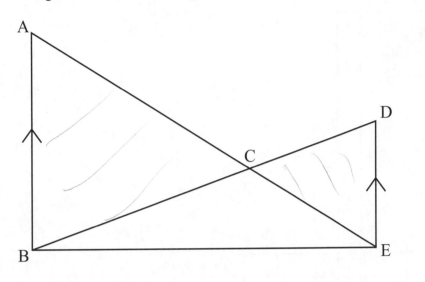

(a)* Prove that triangle ABC is similar to triangle EDC. State your reasons.

....

....

....

....
                                                                                                                        (3 marks)

(b) Explain why you cannot say whether triangles ABE and DEB show congruence.

....

....

....
                                                                                                                        (2 marks)

18   The ages of 190 members of a sports club are shown on the histogram below.

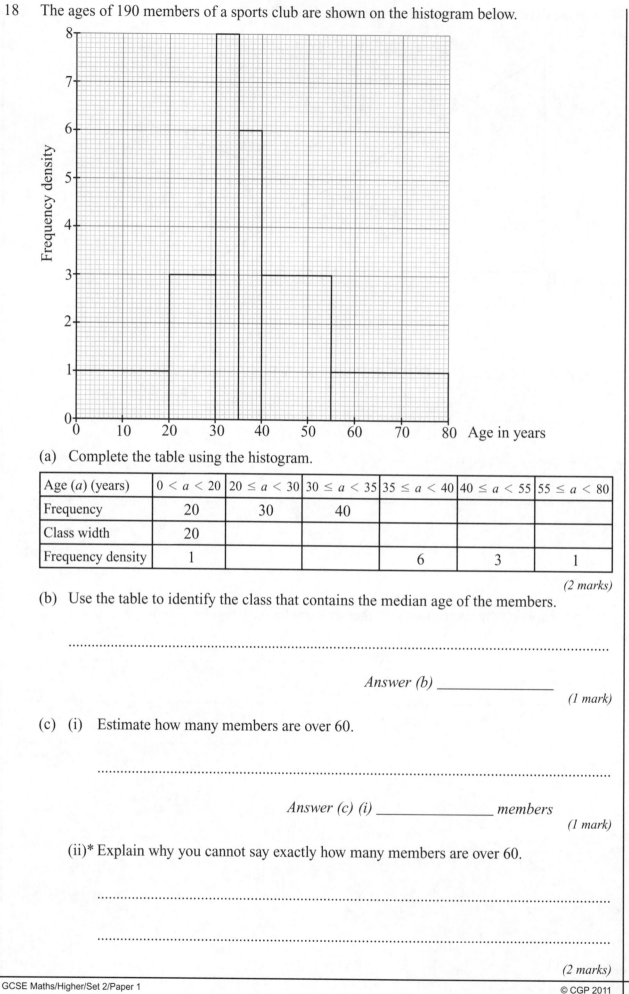

(a) Complete the table using the histogram.

Age (a) (years)	$0 < a < 20$	$20 \leq a < 30$	$30 \leq a < 35$	$35 \leq a < 40$	$40 \leq a < 55$	$55 \leq a < 80$
Frequency	20	30	40			
Class width	20					
Frequency density	1			6	3	1

(2 marks)

(b) Use the table to identify the class that contains the median age of the members.

....

Answer (b) _____

(1 mark)

(c) (i)   Estimate how many members are over 60.

....

Answer (c) (i) _____ *members*

(1 mark)

(ii)* Explain why you cannot say exactly how many members are over 60.

....

....

(2 marks)

19   (a)   Use the axes below to draw the graph of $x^2 + y^2 = 9$.

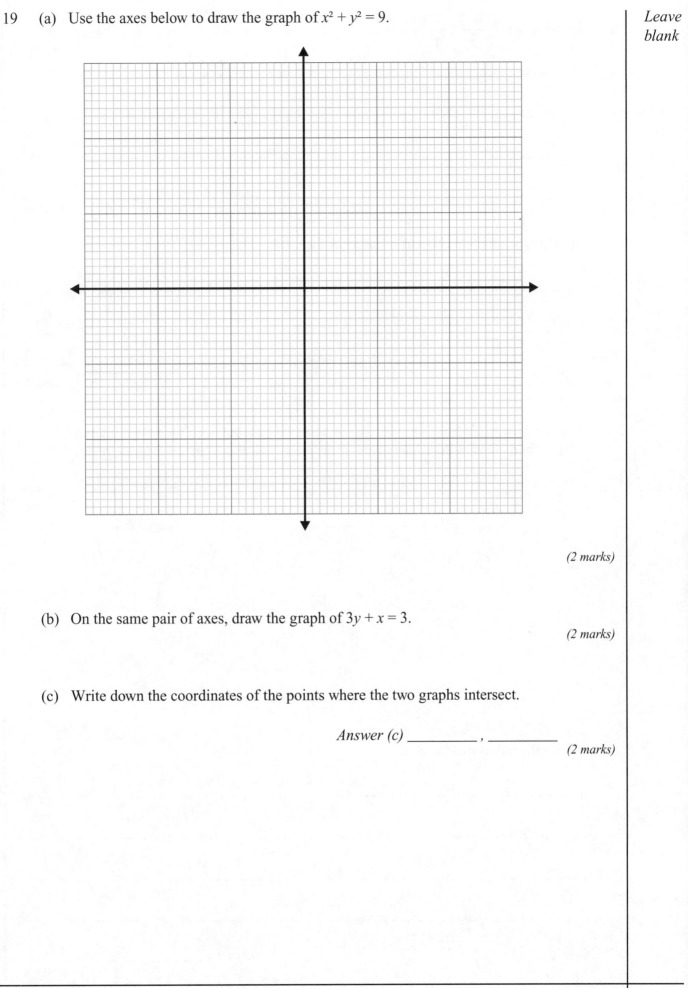

(2 marks)

(b)   On the same pair of axes, draw the graph of $3y + x = 3$.

(2 marks)

(c)   Write down the coordinates of the points where the two graphs intersect.

Answer (c) _____ , _____

(2 marks)

20   (a) Simplify $\dfrac{x^2 + 3x - 10}{x^2 - 4}$.

....

....

....

Answer (a) _____

(3 marks)

(b) Using your answer to part (a) or otherwise, solve $\dfrac{x^2 + 3x - 10}{x^2 - 4} = 2$.

....

....

Answer (b) x = _____

(2 marks)

21   A cone can be made from a circle: part of a circle is cut out, then the remaining part is folded and glued at the edges. This forms a cone, as shown in the diagram below.

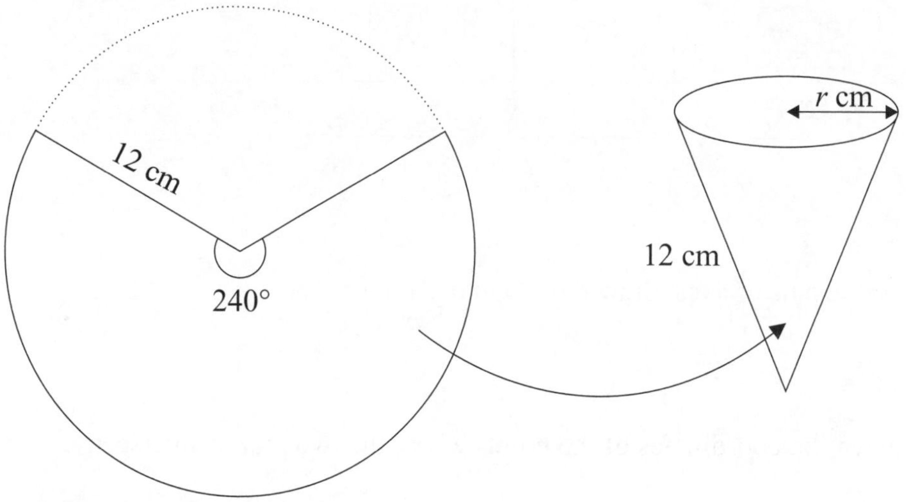

What is the radius *r* of the cone?

....

....

....

....

Answer _____ *cm*

(4 marks)

GCSE Mathematics
Practice Set 2

Paper 2 Calculator

Higher Tier

Time allowed: 1 hour 45 minutes

Centre name				
Centre number				
Candidate number				

Surname
Other names
Candidate signature

In addition to this paper you should have:
- GCSE Mathematics Formula Sheet: Higher Tier.
- A calculator
- A ruler.
- A protractor.
- A pair of compasses.

Tracing paper may be used.

For examiner's use							
Q	Attempt №			Q	Attempt №		
	1	2	3		1	2	3
1				11			
2				12			
3				13			
4				14			
5				15			
6				16			
7				17			
8				18			
9				19			
10				20			
				21			
Total							

Instructions to candidates
- Write your name and other details in the spaces provided above.
- Answer **all** questions in the spaces provided.
- In calculations show clearly how you worked out your answers.
- Take the value of π to be 3.142, or use the π button on your calculator.

Information for candidates
- The marks available are given in brackets at the end of each question.
- You may get marks for method, even if your answer is incorrect.
- In questions labelled with an asterisk *, you will be assessed on the quality of your written communication — take particular care here with spelling, punctuation and the quality of explanations.

Advice to candidates
- Work steadily through the paper.
- Don't spend too long on one question.
- If you have time at the end, go back and check your answers.

1   Jan and Sal are given some money by their grandmother. The amounts she gives them are in the same ratio as their ages. Jan is 8 and gets £36.

    (a) Sal gets £63. How old is she?

....

....

Answer (a) _____

(2 marks)

    (b) Sal decides to split her money. Sal and her boyfriend share it in the ratio 4:3. How much money does Sal give to her boyfriend?

....

....

*Answer (b) £*_____

(2 marks)

2   Solve:

    (a) $4(x + 2) = 16$

....

Answer (a) x = _____

(1 mark)

    (b) $3x - 4 = 6x + 8$

....

....

Answer (b) x = _____

(2 marks)

3   Mary wants to cook a vegetable bake using the ingredients below.
     The recipe comes from an old cookery book that only uses imperial measures.
     Mary's cooking equipment only uses metric measures.

<u>Vegetable bake (serves 4)</u>

1 lb potatoes   2 oz flour
1 lb mixed root vegetables   1 pint milk
(carrots, swede, parsnips)   1 oz butter
12 oz onions   4 oz grated cheese

(a) Calculate how many kilograms of potatoes Mary needs.

     ....

                                        Answer (a) _____ *kg*
                                                                    (1 mark)

(b) Mary has 1 litre of milk. Does she have enough milk to make the bake?
     Explain your answer.

     ....

     ....
                                                                    (1 mark)

(c) The recipe says the bake should be cooked in an oven at 390 °F.
     Mary's oven only shows temperatures in °C.
     She knows the formula to convert from °F to °C is:

$$F = \frac{9C}{5} + 32$$ where C = temperature in °C, F = temperature in °F

     (i) Rearrange this formula to make C the subject.

          ....

          ....

                                        Answer (c) (i) C = _____
                                                                    (1 mark)

     (ii) Estimate the temperature Mary should set her oven to in °C.

          ....

          ....
                                        Answer (c) (ii) _____ *°C*
                                                                    (2 marks)

4   In a survey customers visiting two out of town supermarkets, Rollinsons and Asco, were
     asked why they chose to shop in each supermarket. The table below shows the responses
     of Rollinsons customers to the statement: "This supermarket offers good value for money".

Response	Number of Customers
Agree strongly	70
Mostly agree	80
Neither agree nor disagree	25
Mostly disagree	20
Strongly disagree	5

(a) Draw a pie chart to show this data. Show your working.

(4 marks)

(b)* The pie chart on the right shows the responses
     from Asco customers to the same statement,
     "This supermarket offers good value for money".

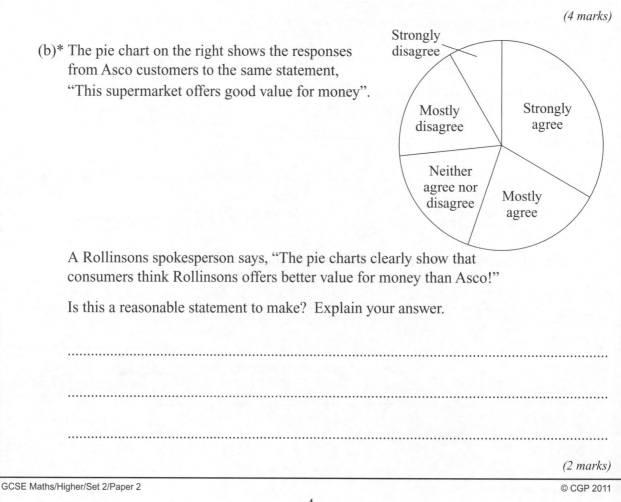

     A Rollinsons spokesperson says, "The pie charts clearly show that
     consumers think Rollinsons offers better value for money than Asco!"

     Is this a reasonable statement to make? Explain your answer.

     ....

     ....

     ....

(2 marks)

5   The grid below shows two shapes, labelled P and Q.

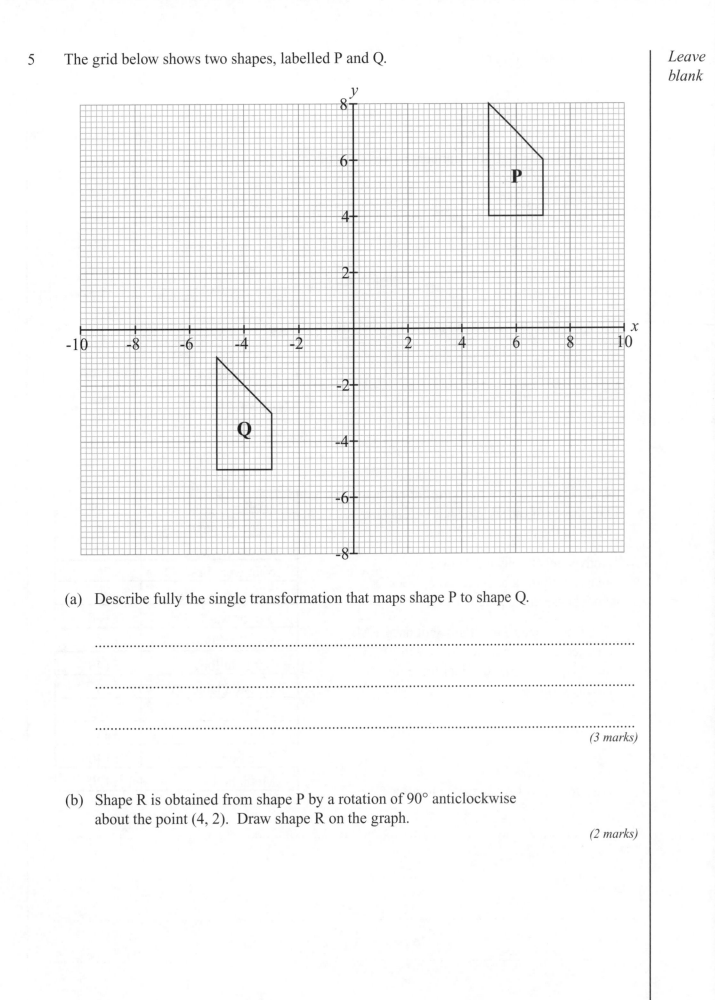

(a) Describe fully the single transformation that maps shape P to shape Q.

....

....

....
                                                                                              (3 marks)

(b) Shape R is obtained from shape P by a rotation of 90° anticlockwise
    about the point (4, 2). Draw shape R on the graph.

                                                                                              (2 marks)

6   (a) Find the value of:

$$\frac{9.87 + \sqrt{24.6}}{0.82^2 \times 6.54}$$

Write down all the numbers on your calculator display.

....

....

Answer (a) _____

(2 marks)

(b) Round your answer to part (a) to 3 significant figures.

....

Answer (b) _____

(1 mark)

7   Andy is buying new furniture for his bedroom. The prices of the style he has chosen are listed in the table on the right.

Andy wants to buy a 3 drawer chest, a large tallboy, a bedside table, a desk, a bookcase and a pine bed.

The final price of the items will have VAT, charged at 20%, added. Andy will get a 20% discount on the final price for paying cash.

Item	Price
3 drawer chest	£48
4 drawer chest	£60
5 drawer chest	£76
2 door wardrobe	£88
3 door wardrobe	£120
Small tallboy	£95
Large tallboy	£140
Bedside table	£28
Desk	£50
Bookcase	£43
Pine bed	£140
Metal bed	£120

How much will Andy pay in total?

....

....

....

....

Answer £ _____

(4 marks)

8   Triangle ABC has side AB = 10 cm, side AC = 8.5 cm and angle BAC = 40°.

Draw and label the triangle ABC accurately. Measure the length of side BC.

Answer BC = _____ *cm*

(3 marks)

9   The table below shows the scores of 57 students in a maths test.

Mark	0-15	16-30	31-45	46-60	61-75	76-90
Frequency	0	13	18	14	9	3

Draw a frequency polygon to show this data.

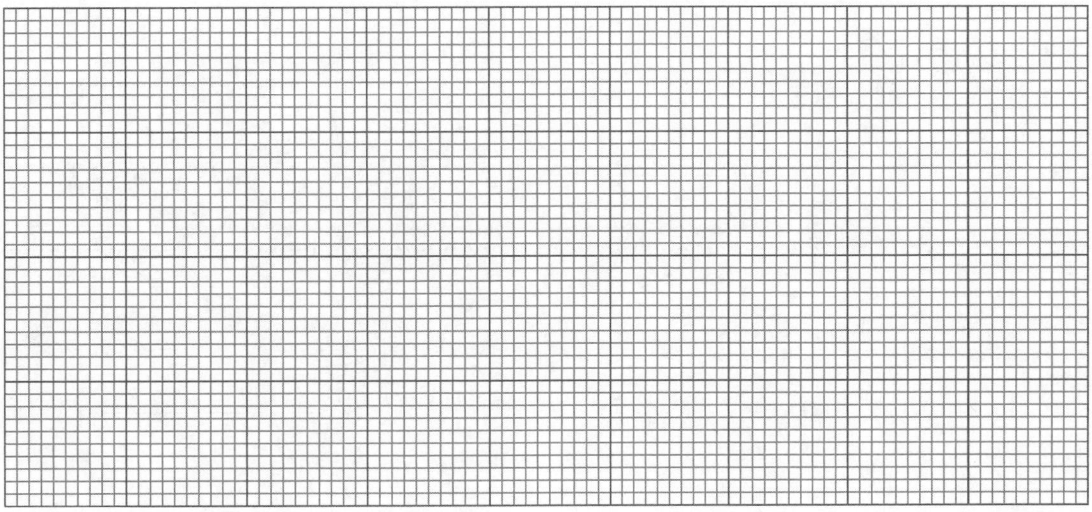

(3 marks)

10 Jordan has made a cake with the dimensions shown.

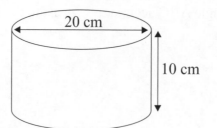

(a) What is the volume of the cake? Give your answer in terms of π.

....

....

Answer (a) _____ cm^3

(2 marks)

(b) The cake weighs 3 kg. What is the average density of the cake in g/cm^3?
Give your answer to 2 decimal places.

....

....

....

Answer (b) _____ g/cm^3

(3 marks)

(c) Jordan wants to cover the top and sides of the cake with a layer of marzipan.
He knows that a 100 g pack of marzipan is enough to cover about 200 cm^2 of cake.
How many packs of marzipan does Jordan need?

....

....

....

....

Answer (c) _____ *packs*

(4 marks)

11 Factorise:

(a) $3x^2y - 9xy^3 + 6x^3$

....

....

Answer (a) _____
(2 marks)

(b) $5x^2 + 6x - 8$

....

....

Answer (b) _____
(2 marks)

12 Ken wants to lose 6 lbs before he goes on holiday in 4 weeks' time. He researches weight loss on the internet and finds that to lose 1 lb, he needs to eat 3500 fewer kCal than he uses.

To lose the weight, he decides to cut out snacks from his diet, and go to one type of exercise class 3 times a week.

The tables below show the average number of kCal in the snacks Ken would normally eat in 4 weeks, and the average number of kCal burned per exercise session.

Snacks eaten in 4 weeks	kCal
Biscuits	1800
Crisps	3200
Chocolate bars	6100
Muffins	1450

Exercise class	kCal burned
Spin class	720
Aerobics	420
Circuit training	700
Fitness class	440

Which exercise class could Ken join to lose the weight in time for his holiday?
Show all your working.

....

....

....

....

Answer _____
(4 marks)

13   Sarah has just inherited £5000.
      She wants to invest her money for 2 years in one of the following accounts:

      • Account 1: Pays 3.5% compound interest per annum.

      • Account 2: Pays 5% compound interest per annum.
         One-off fee of £150 to be paid when the account is opened.

      Which of these accounts should Sarah invest in?
      How much more interest will she make by choosing that account?

....

....

....

....

....
                                                                                            (4 marks)

14

                                                              Not to scale.

      (a) Calculate the size of angle A.

....

....

....

                                    Answer (a) _____ °
                                                                          (3 marks)

      (b) Calculate the area of the triangle. Give your answer to the nearest cm².

....

....

                                    Answer (b) _____ *cm²*
                                                                          (2 marks)

15   Sandy is doing an experiment to investigate the growth of seedling plants. He has grown two trays of seedlings, Tray A and Tray B, and recorded their heights in the frequency tables below.

Tray A

Height in mm	Frequency
$0 < h \le 5$	9
$5 < h \le 10$	19
$10 < h \le 15$	28
$15 < h \le 20$	34
$20 < h \le 25$	19
$25 < h \le 30$	11

Tray B

Height in mm	Frequency
$0 < h \le 5$	2
$5 < h \le 10$	8
$10 < h \le 15$	50
$15 < h \le 20$	47
$20 < h \le 25$	11
$25 < h \le 30$	2

(a)   Estimate the mean height of seedlings in Tray A. Show your working.

....

....

....

....

....

Answer (a) _____ *mm*

(4 marks)

The cumulative frequency curves for Trays A and B are drawn on the graph below.

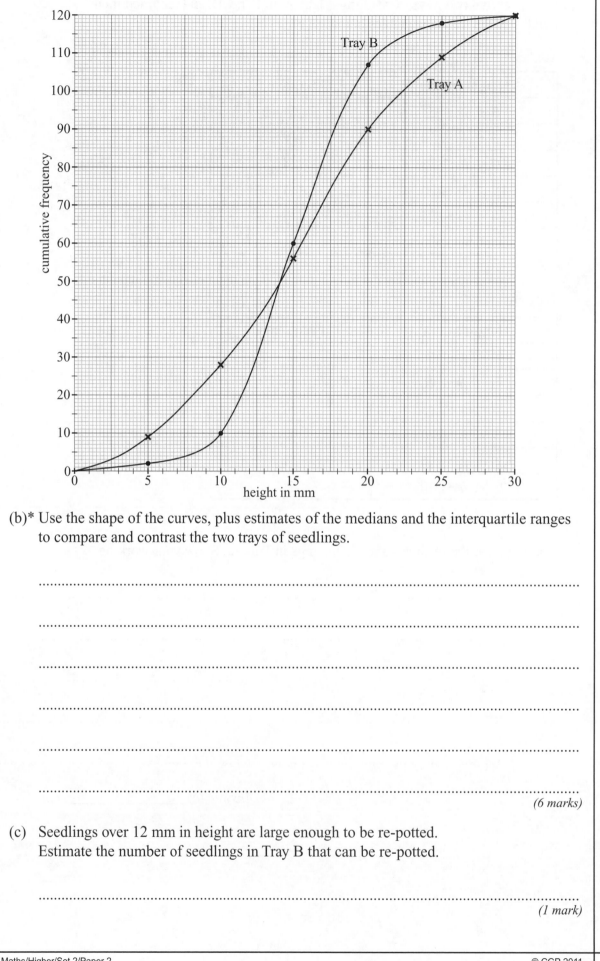

(b)* Use the shape of the curves, plus estimates of the medians and the interquartile ranges to compare and contrast the two trays of seedlings.

....

....

....

....

....

....

(6 marks)

(c) Seedlings over 12 mm in height are large enough to be re-potted.
Estimate the number of seedlings in Tray B that can be re-potted.

....

(1 mark)

16   A rectangle is $2x$ cm long and $(x - 5)$ cm wide, with an area of 48 cm².

Find the value of x.

....

....

....

....

....

....

Answer $x =$ _____

(5 marks)

17*   A circular paddling pool has a surface area of 2 m², to the nearest 0.1 m².
The maximum depth of water the pool can hold is 20 cm, to the nearest 2 cm.
The bucket used to fill the pool holds 10 litres, to the nearest 0.5 litres.

Can you be certain that if you put 35 full buckets of water in the pool,
you will completely fill the pool? Show all your working and explain your answer.

....

....

....

....

....

....

....

....

(5 marks)

18   Eleanor wants to calculate the acceleration and final speed of a toy car rolling down a ramp. Leave blank
      At the top of the ramp, the car has a starting speed of 0.2 m/s.
      It takes the car 2.3 seconds to travel a distance of 1.5 m to the end of the ramp.

      Eleanor has a list of formulae she can use to help her:

$$v^2 = u^2 + 2as \qquad\qquad a = \text{acceleration}$$
$$v = u + at \qquad\qquad\quad s = \text{distance travelled}$$
$$s = ut + \tfrac{1}{2}at^2 \qquad\qquad t = \text{time taken}$$
$$F = ma \qquad\qquad\qquad u = \text{starting speed}$$
$$(v + u)/2 = s/t \qquad\qquad v = \text{final speed}$$

      Calculate the final speed and acceleration of the car at the end of the ramp.
      Show your working and state the units of your answers.

      ....

      ....

      ....

      ....

      ....

      ....

      Answer v = _____   *Answer a =* _____

                                                                          (5 marks)

19   Simplify:

      (a) $3n^3m^5 \times 4n^{-6}m^4$

            ....

            ....

                                                    Answer (a) _____

                                                                          (2 marks)

      (b) $6p^{-2}q^{-3} \div 3p^{-6}q^4$

            ....

            ....

                                                    Answer (b) _____

                                                                          (2 marks)

20 On the cuboid below, AB measures 12 cm, AD measures 9 cm and BF measures 6 cm.

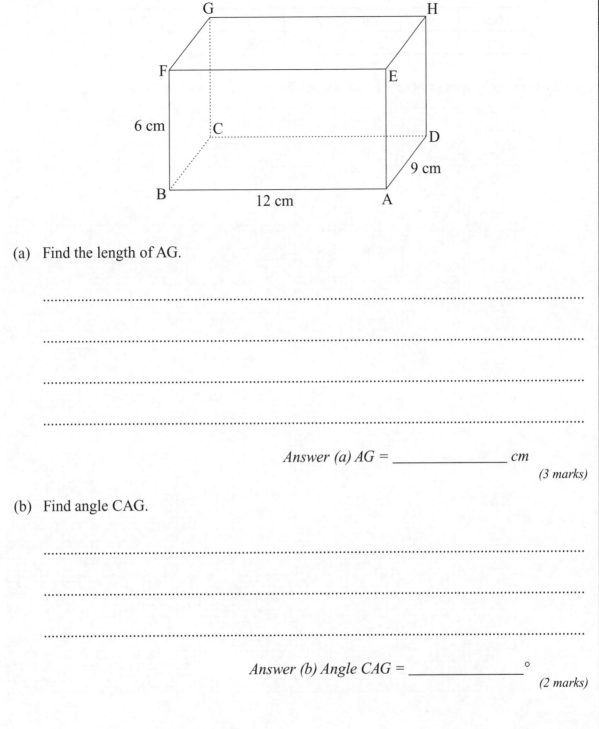

(a) Find the length of AG.

....

....

....

....

Answer (a) AG = _____ cm

(3 marks)

(b) Find angle CAG.

....

....

....

Answer (b) Angle CAG = _____ °

(2 marks)

21   Charlotte has measured displacement (d) and mass (m) in a physics experiment.
     Her results are shown in the table below.

d (cm)	0	1.8	2.5	3.1	3.6	4.0	4.4
m (g)	0	300	600	900	1200	1500	1800

By drawing graphs or otherwise, work out if $d \propto m$, $d^2 \propto m$ or $d^3 \propto m$.

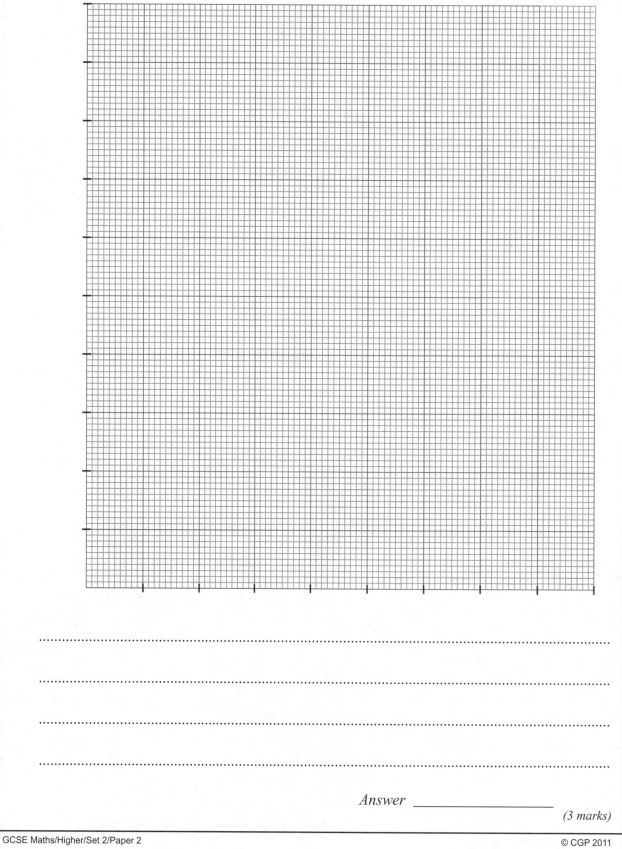

....

....

....

....

Answer _____

(3 marks)

General Certificate of Secondary Education

GCSE Mathematics

Practice Set 3
Paper 1 Non-calculator

Higher Tier

Time allowed: 1 hour 45 minutes

Centre name				
Centre number				
Candidate number				

Surname	
Other names	
Candidate signature	

In addition to this paper you should have:
- GCSE Mathematics Formula Sheet: Higher Tier.
- A ruler.
- A protractor.
- A pair of compasses.

Tracing paper may be used.

For examiner's use							
Q	Attempt Nº			Q	Attempt Nº		
	1	2	3		1	2	3
1				12			
2				13			
3				14			
4				15			
5				16			
6				17			
7				18			
8				19			
9				20			
10				21			
11				22			
Total							

Instructions to candidates
- Write your name and other details in the spaces provided above.
- Answer **all** questions in the spaces provided.
- In calculations show clearly how you worked out your answers.

Information for candidates
- The marks available are given in brackets at the end of each question.
- You may get marks for method, even if your answer is incorrect.
- In questions labelled with an asterisk *, you will be assessed on the quality of your written communication — take particular care here with spelling, punctuation and the quality of explanations.

Advice to candidates
- Work steadily through the paper.
- Don't spend too long on one question.
- If you have time at the end, go back and check your answers.

1 Mike is a landscape gardener. Jim has asked Mike to design his garden for him.
Mike has measured and drawn a sketch plan of Jim's garden, shown below.
He has also marked on when different parts of the garden get the sun.

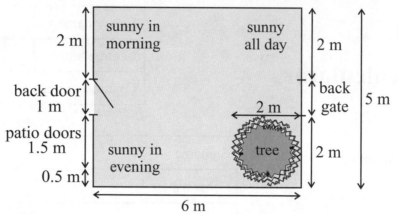

Jim wants:

(i)   a vegetable plot, at least 6 m² in area, which needs as much sun as possible;

(ii) a square patio, which Jim wants to use mainly for barbecues in the evening.
      The patio needs to be at least 2 m²;

(iii) a path 1 m wide to get from the back door of the house to the gate.

In the space below, use a ruler to draw an accurate scale plan that
Mike might draw to plan Jim's garden. Clearly label the features.

(4 marks)

Leave blank

2 This is a regular hexagon.

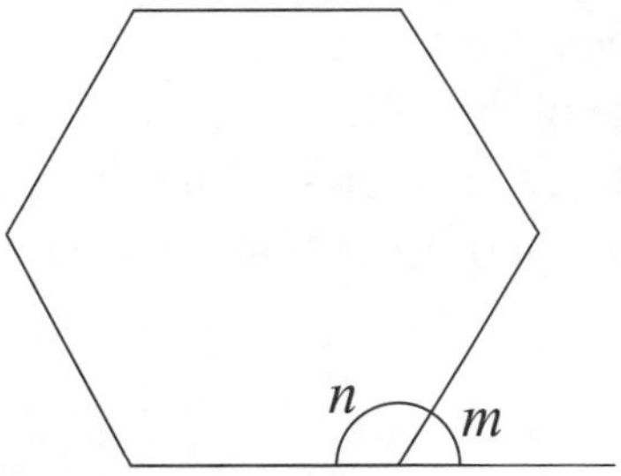

not drawn accurately
n
m

(a) Calculate the size of angle *m*. Show all your working.

....

....

Answer (a) m = ______________ °

(2 marks)

(b) Calculate the size of angle *n*. Show all your working.

....

....

Answer (b) n = ______________ °

(2 marks)

3 Class 11B have done a survey of Years 9, 10 and 11.
They asked 30 pupils from each year group about their preferred fast food.

They began to record the results in the two-way table shown below.

Favourite fast food	**Year 9**	**Year 10**	**Year 11**	**Total**
Burger	12	9	8	29
Fish and chips	10	6	2	
Curry	0			15
Pizza			10	28
Total	**30**	**30**	**30**	**90**

(a) Complete the table to show all the results from the survey.

(2 marks)

(b) There are 120 pupils in Year 9.
Estimate how many pupils in Year 9 prefer fish and chips to other fast foods.

....

....

Answer (b) ______________ *pupils*

(2 marks)

4   Mia is interviewing six people for a job.

All the interviews must take place within one 6 hour day. Only one person can be interviewed at a time, and there must be a 15 minute gap between each interview. Work out the maximum amount of time Mia can spend interviewing each person.

....

....

....

....

Answer _____ *minutes*

(3 marks)

5   Four friends put some money together and buy seven lottery tickets.
     Sheena puts in 80p, Darryl puts in £2, Helen puts in £1.60 and Ali puts in £1.20.

     One of the lottery tickets wins them £42. They decide the fairest way to split the money is to divide the winnings in proportion to the amount of money they put in.

     (a) Write down the amount of money each person contributed as a
          ratio in its simplest form.

....

....

Answer (a) _____

(1 mark)

     (b) Calculate the amount of money that each of the friends will get.

....

....

....

....

....

....

(4 marks)

6   (a) Express 48 as a product of its prime factors.

....

....

Answer (a) _____

(2 marks)

(b) Find the Lowest Common Multiple of 7 and 12.

....

....

Answer (b) _____

(2 marks)

(c) Find the Highest Common Factor of 35 and 63.

....

....

Answer (c) _____

(2 marks)

7   The diagram below shows the road between two towns, Edgeville (E) and Farrowton (F).
     A new road is planned, which will be the perpendicular bisector of the current road.

     Draw where the new road will be built. Show all your construction lines clearly.

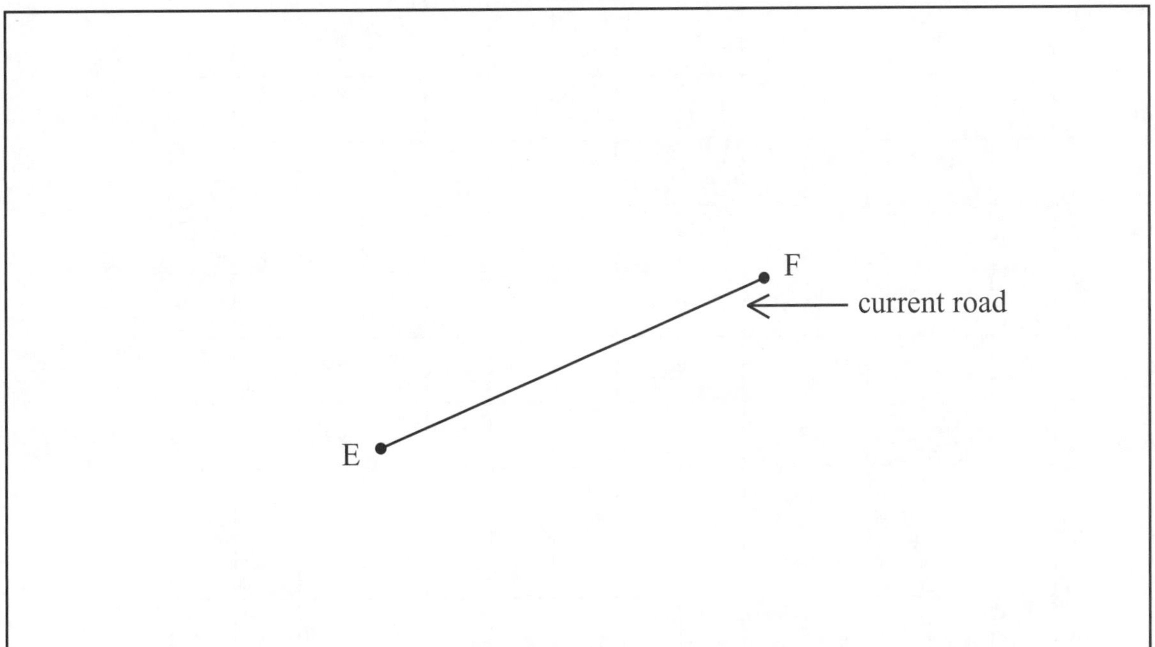

(2 marks)

8   (a) Factorise $6pq + 9p^2 - 24pq^2$.

....

Answer (a) _____

(2 marks)

(b) Expand and simplify $(2j - 7k)(3j + 4k)$.

....

....

Answer (b) _____

(2 marks)

(c) Expand $(2x - 3xy)^2$.

....

....

Answer (c) _____

(2 marks)

9   Enlarge Shape A on the grid below using a scale factor of -2,
     and a centre of enlargement (5, 4).

(3 marks)

10 (a) Convert 6500 mm² into cm².

....

....

Answer (a) _____ *cm²*

(1 mark)

(b) A cube has a volume of 8 m³. Convert the volume of the cube into cm³.

....

....

....

Answer (b) _____ *cm³*

(2 marks)

11 (a) Adlai is designing a questionnaire to give to shoppers outside a supermarket. The first question in the questionnaire is shown below.

> *Questionnaire for Shoppers*
>
> 1)   How old are you? Tick a box below.
>
>   0 to 18 ☐   18 to 30 ☐   30 to 50 ☐   50+ ☐

(i) Give one criticism of the way the question above has been asked.

....

(1 mark)

(ii) Suggest one way in which the question could be improved.

....

(1 mark)

(b) The aim of the questionnaire is to find out how much money customers spend in the supermarket. Design a suitable question that could be used in the questionnaire.

....

....

....

(2 marks)

12   An electricity company charges the following rates:

Peak: 29.14 p/unit
Off-peak: 10.48 p/unit

Mrs Jones receives an electricity bill, shown below.

Electricity bill for: Mrs Jones

Charges: Units Used:

Peak: 29.14 p/unit   = 162 units

Off-peak: 10.48 p/unit   = 146 units

Sub-total of charges before VAT   = £67.51

VAT (5%)   = £3.38

Total charges including VAT   = £70.89

Mrs Jones thinks that either her bill or VAT has been calculated incorrectly,
as the final amount seems too expensive.

Round the numbers to work out an estimate for the correct value of the bill, including VAT.

....

....

....

....

....

....

....

Answer £ _____

(5 marks)

13   Look at the shapes *A* and *B* on the grid below.

(a) Describe fully the single transformation that maps shape 'A' to shape 'B'.

....

....

....
(3 marks)

(b) Draw and label the line $y = -x$ on the grid above.
(1 mark)

(c) Draw the reflection of shape 'A' in the line $y = -x$. Label this shape 'C'.
(1 mark)

14   The table below shows the wages of 50 students with part-time jobs.

Hourly Wage (W)	Frequency (F)	Cumulative Frequency (CF)
£4.50 ≤ W < £5.00	5	
£5.00 ≤ W < £5.50	12	
£5.50 ≤ W < £6.00	15	
£6.00 ≤ W < £6.50	8	
£6.50 ≤ W < £7.00	5	
£7.00 ≤ W < £7.50	3	
£7.50 ≤ W < £8.00	2	

(a) Complete the cumulative frequency column of the table.

(2 marks)

(b) Draw a cumulative frequency graph of this data on the axes below.

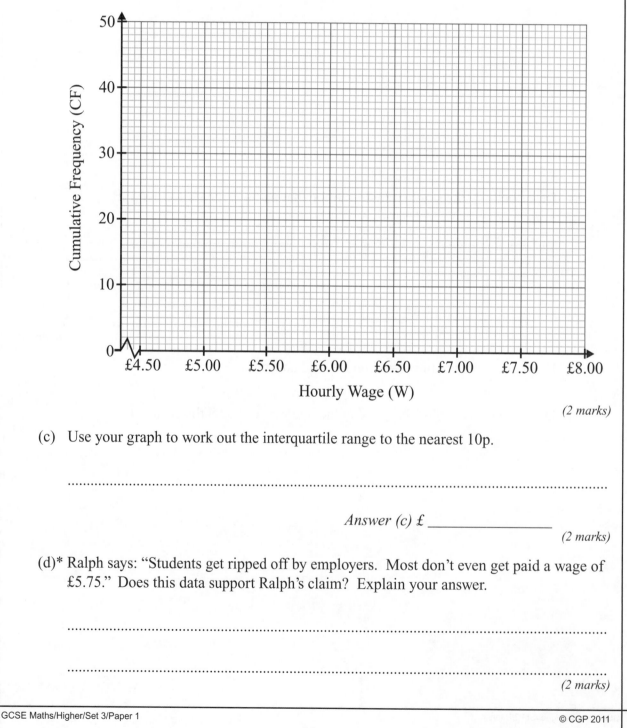

(2 marks)

(c) Use your graph to work out the interquartile range to the nearest 10p.

....

Answer (c) £ _____

(2 marks)

(d)* Ralph says: "Students get ripped off by employers. Most don't even get paid a wage of £5.75." Does this data support Ralph's claim? Explain your answer.

....

....

(2 marks)

15   The rectangle below has a length of $(x + 4)$ cm.

$(x + 4)$ cm   *not drawn accurately*

(a)   The width of the rectangle is exactly 5 cm less than the length.
Write down and simplify an expression for the width of the rectangle.

....

Answer (a) _____ *cm*

(1 mark)

(b)   The area of the rectangle is 14 cm². Use this information to find
the perimeter of the rectangle. You must show all your working.

....

....

....

....

Answer (b) _____ *cm*

(4 marks)

16   Petra wants to buy a new car, keep it for 2 years and then sell it on.
Her two favourite types of car are:

•   An Alta Bravio, which costs £9500 new but its value depreciates by 20% each year.

•   A MW Ringo, which costs £9600 new and will be worth £6000 after 2 years.

Petra wants to buy the car that will have the most value after 2 years.
Which car should she buy?

....

....

....

....

Answer _____

(3 marks)

17   Mr Smith rolls a fair dice to decide how long to walk each day.
      If the dice lands on a five, he will take his dog for a 5 mile walk.
      Otherwise he will just take his dog on a 1 mile walk.

    (a)   Draw a probability tree diagram to show all the possible outcomes for 2 days.

                                                                        (2 marks)

    (b)   Work out the probability that Mr Smith walks his dog 5 miles on both days.

          ....

          ....

                                        Answer (b) _____
                                                                        (2 marks)

    (c)   Work out the probability that Mr Smith will go for a 5 mile walk on at least one day.

          ....

          ....

                                        Answer (c) _____
                                                                        (2 marks)

18   (a)   Simplify $\sqrt{20} + \sqrt{45}$

....©CGP.....

....

....

....

Answer (a) _____
                                          (2 marks)

(b)   Rationalise the denominator of $\dfrac{1}{\sqrt{3}}$

....

....

....

....

Answer (b) _____
                                          (2 marks)

(c)   Rationalise the denominator of $\dfrac{2 + \sqrt{3}}{2 - \sqrt{3}}$

....

....

....

....

Answer (c) _____
                                          (2 marks)

19   Evaluate the expressions given below. Write your answers in their simplest form.

(a) $8^{\frac{2}{3}}$

....

....

Answer (a) _____

(2 marks)

(b) 4^{-3}

....

....

Answer (b) _____

(2 marks)

(c) $\left(\frac{27}{8}\right)^{-\frac{4}{3}}$

....

....

....

Answer (c) _____

(3 marks)

20   The graph $y = \sin x$ for values of x between $0°$ and $360°$ is shown below.

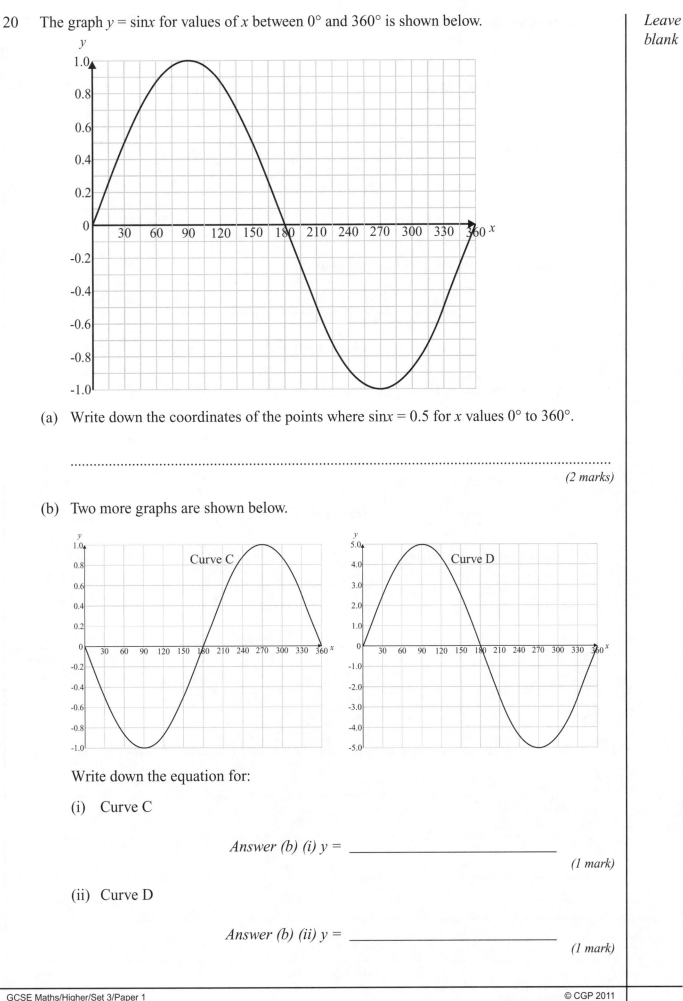

(a) Write down the coordinates of the points where $\sin x = 0.5$ for x values $0°$ to $360°$.

....

(2 marks)

(b) Two more graphs are shown below.

Write down the equation for:

(i)   Curve C

Answer (b) (i) y = _____

(1 mark)

(ii) Curve D

Answer (b) (ii) y = _____

(1 mark)

21   The diagram shows a triangle MNQ. MR = 2RN and QS = 2SN.

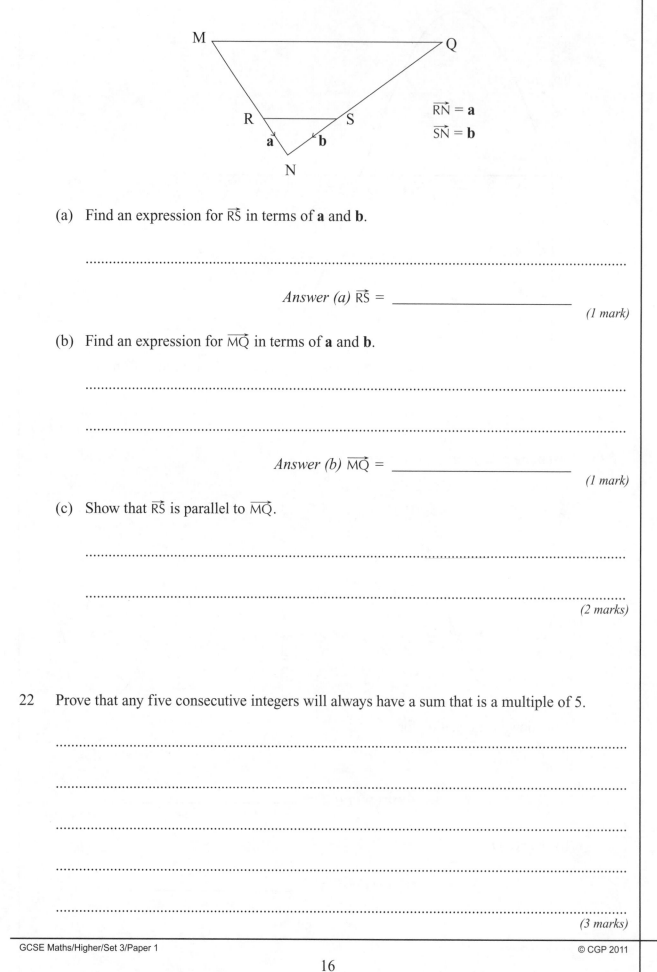

$$\overrightarrow{RN} = \mathbf{a}$$
$$\overrightarrow{SN} = \mathbf{b}$$

(a) Find an expression for $\overrightarrow{RS}$ in terms of **a** and **b**.

....

Answer (a) $\overrightarrow{RS}$ = _____

(1 mark)

(b) Find an expression for $\overrightarrow{MQ}$ in terms of **a** and **b**.

....

....

Answer (b) $\overrightarrow{MQ}$ = _____

(1 mark)

(c) Show that $\overrightarrow{RS}$ is parallel to $\overrightarrow{MQ}$.

....

....

(2 marks)

22   Prove that any five consecutive integers will always have a sum that is a multiple of 5.

....

....

....

....

....

(3 marks)

General Certificate of Secondary Education

GCSE Mathematics

Practice Set 3

Paper 2 Calculator

Higher Tier

Time allowed: 1 hour 45 minutes

Centre name				
Centre number				
Candidate number				

Surname	
Other names	
Candidate signature	

In addition to this paper you should have:
- GCSE Mathematics Formula Sheet: Higher Tier.
- A calculator.
- A ruler.
- A protractor.
- A pair of compasses.

Tracing paper may be used.

For examiner's use							
Q	Attempt Nº			Q	Attempt Nº		
	1	2	3		1	2	3
1				11			
2				12			
3				13			
4				14			
5				15			
6				16			
7				17			
8				18			
9				19			
10				20			
Total							

Instructions to candidates
- Write your name and other details in the spaces provided above.
- Answer **all** questions in the spaces provided.
- In calculations show clearly how you worked out your answers.
- Take the value of π to be 3.142, or use the π button on your calculator.

Information for candidates
- The marks available are given in brackets at the end of each question.
- You may get marks for method, even if your answer is incorrect.
- In questions labelled with an asterisk *, you will be assessed on the quality of your written communication — take particular care here with spelling, punctuation and the quality of explanations.

Advice to candidates
- Work steadily through the paper.
- Don't spend too long on one question.
- If you have time at the end, go back and check your answers.

1   (a)   Write $\frac{6}{32}$ as a decimal.

....

Answer (a) _____

(1 mark)

(b)   Write $\frac{6}{9}$ as a recurring decimal.

....

Answer (b) _____

(1 mark)

(c)   Write $0.1\dot{8}$ as a fraction in its simplest form.

....

Answer (c) _____

(2 marks)

2   Solve:

(a) $4p + 8 = 36$

....

....

Answer (a) p = _____

(1 mark)

(b) $6q + 13 = 12q - 2$

....

....

Answer (b) q = _____

(2 marks)

(c) $3(2r + 14) = 2(25 + 4r)$

....

....

Answer (c) r = _____

(3 marks)

3   (a)   Complete the table of values below for the graph $y = x^2 - 2x + 1$.

x	-1	0	1	2	3
y	4				

(2 marks)

(b) Use the table above to draw the graph of $y = x^2 - 2x + 1$ on the axes below.

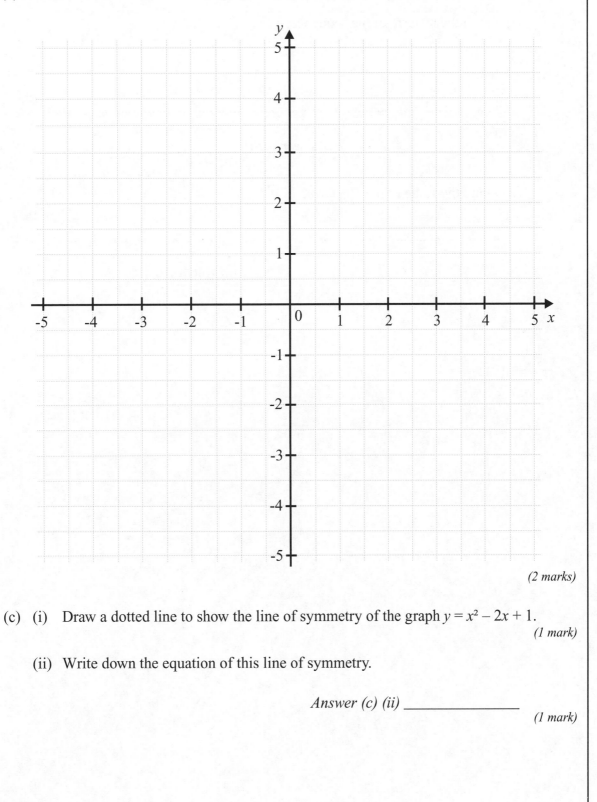

(2 marks)

(c)   (i)   Draw a dotted line to show the line of symmetry of the graph $y = x^2 - 2x + 1$.

(1 mark)

(ii)   Write down the equation of this line of symmetry.

Answer (c) (ii) _____

(1 mark)

4   (a)   A sequence has the nth term $3n + 1$. Write down the first 5 terms of the sequence.

....

....

(2 marks)

(b)   Another sequence begins; 8, 13, 18, 23, 28...

(i)   Find the nth term of the sequence.

....

....

Answer (b)(i) _____

(2 marks)

(ii)   Find the value of the 50th term.

....

Answer (b)(ii) _____

(1 mark)

5   Sammy is training for a 400 m race. She times how long it takes to run 400 m to the nearest second. She has recorded her results for her last 20 tries as a list. She has begun to put her results into a frequency table.

(a) Use Sammy's list to complete the frequency table.

~~62,~~   ~~62,~~   ~~52,~~   ~~70,~~   ~~65,~~   64,   63,   65,   68,   64,

56,   57,   64,   67,   72,   63,   57,   62,   55,   71

Time in seconds	Tally	Frequency
48 – 52	I	
53 – 57		
58 – 62	II	
63 – 67	I	
68 – 72	I	
Total		

(2 marks)

(b) She wants to know if she has improved her average time from last month. Estimate the mean time it takes her to run 400 m. Show your working.

....

....

....

....

....

....

....

Answer (b) _____ *seconds*

(4 marks)

6   Reena works as a Sales Assistant in a furniture shop. The shop is having
     a '15% day' where all its goods, including sale items, are reduced by 15%.
     Reena is given the task of labelling all items with their reduced prices.

   (a)   The current price of a floating shelf is £27.50.
         What price should Reena label the shelf? Give your answer to the nearest penny.

         ....

         ....

                                                    Answer (a) £_____
                                                                        (2 marks)

   (b)   The price of a side table has already been reduced by 25% from its original price
         of £50. What price should Reena label the side table on the '15% day'?

         ....

         ....

         ....

                                                    Answer (b) £_____
                                                                        (3 marks)

7   An outdoor pursuits centre is designing a new obstacle course.

     As part of the course, they want to string a climbing rope diagonally
     between two trees 7.5 m apart as shown in the diagram.

     There needs to be a 70 cm allowance of rope
     at either end to safely attach the rope to the trees.

     How long a piece of rope do they need?

     Give your answer to the nearest 10 cm.

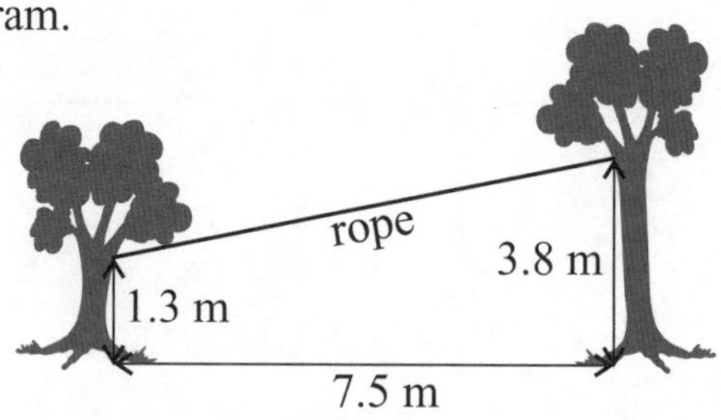

     rope   3.8 m
     1.3 m
     7.5 m

     ....

     ....

     ....

     ....

     ....

                                            Answer _____ *m*
                                                                (4 marks)

8   A triangle has a height of $5x$ and a base length of $7x$.
     The area of the triangle is 63 cm².

(a)   Write an equation in its simplest form to show this information.

     ....

     ....

                              Answer (a) _____

                                                                    (2 marks)

(b) Find the height of the triangle.

     ....

     ....

                              Answer (b) _____ *cm*
                                                                    (3 marks)

9   Leo works in a Perfume Sales Department earning £7.25 per hour.
     He works eight hours a day.
     Leo receives a bonus payment of £1.50 every time he sells *Brand X* perfume.
     His manager notes that Leo has made four *Brand X* sales in the past 60 working days.

(a) Estimate the probability that Leo will not make a *Brand X* perfume sale tomorrow.
     Give your answer as a fraction in its simplest form.

     ....

     ....

                              Answer (a) _____
                                                                    (2 marks)

(b) Leo asks his manager for an advance on his next month's wage. His manager agrees.
     There are 22 working days in the next month. Work out the amount of money that
     Leo's manager should pay him in advance, including any bonus payments.

     ....

     ....

     ....

                              Answer (b) £_____
                                                                    (3 marks)

10   The information given below relates to the ages of members of a chess club.

Youngest Member = 9
Range = 65
Lower Quartile = 44
Median = 60
Upper Quartile = 66

(a) Draw a box plot to show this information below.

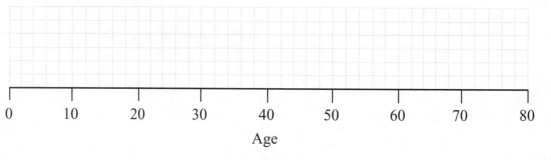

0   10   20   30   40   50   60   70   80

Age

(3 marks)

(b)* The box plot below shows the age distribution of members of a bridge club.

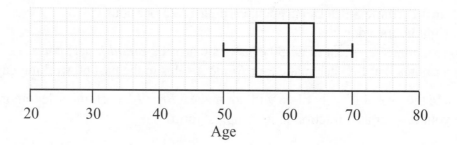

20   30   40   50   60   70   80

Age

Compare the age distributions of the members of the bridge and chess clubs.

....

....

....

....

....

(3 marks)

11   Match each of the graphs below to the correct equation.

A: $y = x$ B: $y = x + 2$ C: $x + y = 5$ D: $y = x^2$

E: $y = -x$ F: $y = x^3$ G: $y = 2$ H: $x = 2$

(a)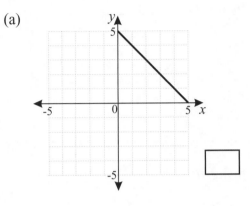

(b)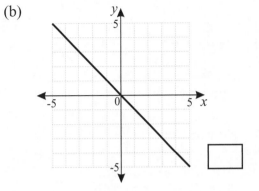

(c)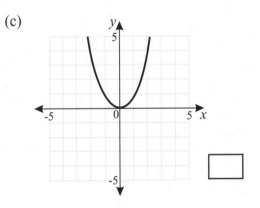

(d)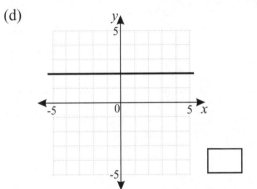

(e)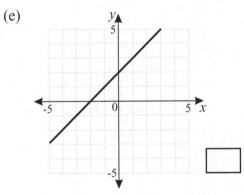

(f)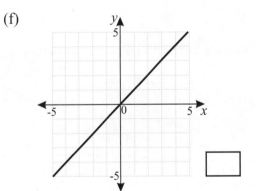

(g)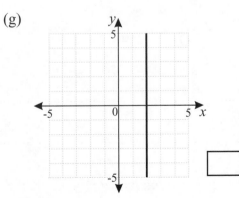

(h)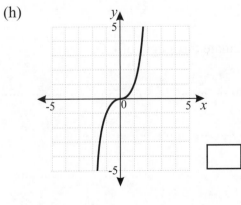

(4 marks)

12   Make x the subject of each of the formulae below.

(a) $y = x - 3$

....

Answer (a) x = _____
(1 mark)

(b) $2x = y - x$

....

....

Answer (b) x = _____
(2 marks)

(c) $y = 6 - x^2$

....

....

Answer (c) x = _____
(2 marks)

13   Triangles DEF and DFG are similar.

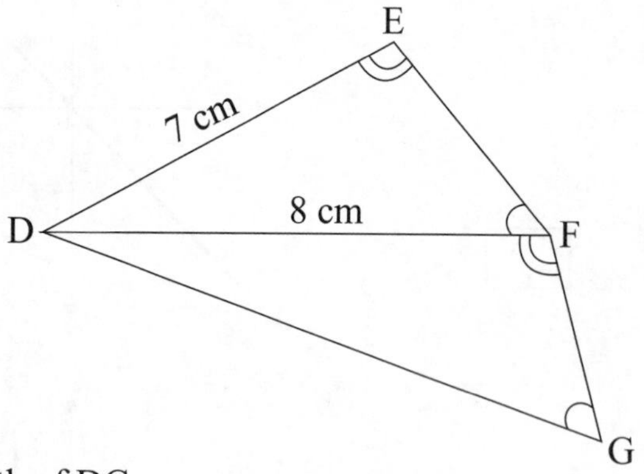

Not to scale.

Calculate the length of DG.

....

....

....

Answer _____ *cm*
(2 marks)

14* (a) Give one reason why it is often preferable to collect data from a sample of people rather than an entire population.

....

....

....

....

(1 mark)

(b) The table below shows the number of pupils within each year group in a school.

Year Group	Number of Pupils
7	180
8	234
9	225
10	173
11	188
Total	**1000**

The Governors of the school are considering changing the school uniform.
They decide to conduct a stratified sample of the pupils in the school
in order to get their opinions. They decide to ask 50 pupils.

Calculate how many pupils should be asked from each year group.

....

....

....

....

....

(3 marks)

15   The time taken to drive along a stretch of motorway is inversely proportional
      to the speed of the vehicle.

      (a) A car travelling at 70 mph covers a distance in 48 minutes. How long will
           it take to cover the same distance driving at 65 mph? Show your working.

           ....

           ....

           ....

                                        Answer (a) _____ *minutes*
                                                                    (3 marks)

      (b) The braking distance of a car travelling is proportional to the speed of the car squared.

           (i)   Write an equation to show this relationship.

                 ....
                                                                    (1 mark)

           (ii) What will happen to a car's braking distance if the speed of the car doubles?

                 ....
                                                                    (1 mark)

16

Not drawn accurately.

      (a) Work out the arc length ST.

           ....

           ....

                                        Answer (a) _____ *cm*
                                                                    (2 marks)

      (b) Work out the area of sector RST. Give your answer to two decimal places.

           ....

           ....

                                        Answer (b) _____ *cm²*
                                                                    (2 marks)

17   Solve the quadratic equations given below.

(a) $x^2 + 9x - 22 = 0$

....

....

Answer (a) x = _____

or x = _____

(2 marks)

(b) $6x^2 + 2x - 4 = 0$

....

....

Answer (b) x = _____

or x = _____

(2 marks)

(c) $3x^2 = x + 7$. Give your answer to 3 significant figures.

....

....

....

....

....

....

Answer (c) x = _____

or x = _____

(3 marks)

18   The table and histogram shown below are both incomplete.
     They show the times taken by 60 students to complete a 2-mile cross-country run.

Time (*m*) minutes	Frequency
$10 < m \leq 13$	
$13 < m \leq 15$	13
$15 < m \leq 16$	
$16 < m \leq 18$	
$18 < m \leq 20$	10
$20 < m \leq 25$	6

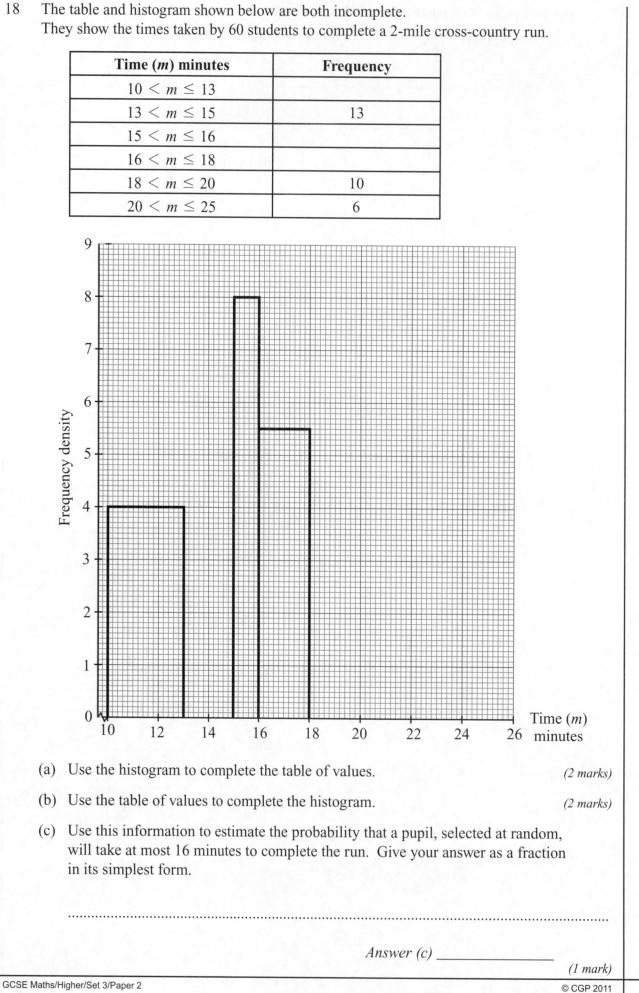

(a) Use the histogram to complete the table of values.

(2 marks)

(b) Use the table of values to complete the histogram.

(2 marks)

(c) Use this information to estimate the probability that a pupil, selected at random,
     will take at most 16 minutes to complete the run. Give your answer as a fraction
     in its simplest form.

....

Answer (c) _____

(1 mark)

19 The width of a door is measured as 75.2 cm to the nearest mm. Leave blank

(a) Write down the smallest possible width of the door.

....

Answer (a) _____ *cm*

(1 mark)

(b) Write down the upper bound for the width of the door.

....

Answer (b) _____ *cm*

(1 mark)

The area of the door is 1.66 m² correct to two decimal places.

(c) Calculate the maximum possible height of the door to the nearest centimetre.

....

....

Answer (c) _____ *m*

(2 marks)

(d) Suggest a value rounded to a suitable degree of accuracy for the height of the door. Give a reason for your answer.

....

....

....

....

(2 marks)

20   The cone, N, shown below has a base radius of 9 cm and a perpendicular height of 15 cm.

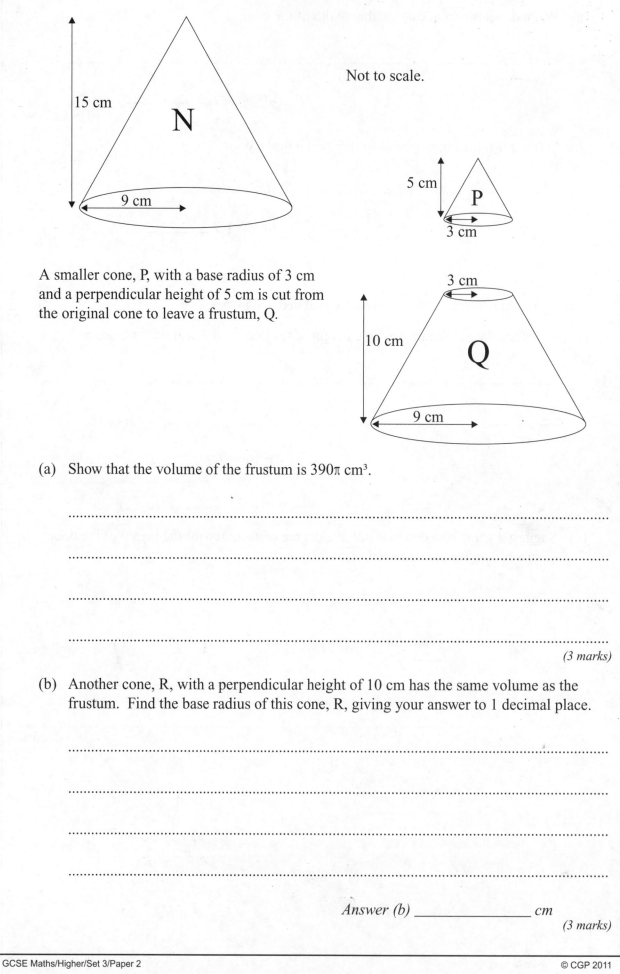

Not to scale.

A smaller cone, P, with a base radius of 3 cm and a perpendicular height of 5 cm is cut from the original cone to leave a frustum, Q.

(a)   Show that the volume of the frustum is 390π cm³.

....

....

....

....

(3 marks)

(b)   Another cone, R, with a perpendicular height of 10 cm has the same volume as the frustum. Find the base radius of this cone, R, giving your answer to 1 decimal place.

....

....

....

....

Answer (b) _____ *cm*

(3 marks)